Chemical Bonds
and Bond Energy

This is Volume 21 in
PHYSICAL CHEMISTRY
A series of monographs
Edited by ERNEST M. LOEBL, *Polytechnic Institue of Brooklyn*

A complete list of the books in this series appears at the end of the volume.

CHEMICAL BONDS AND BOND ENERGY

R. T. SANDERSON

Arizona State University
Tempe, Arizona

A C A D E M I C P R E S S New York and London 1971

124039

ACADEMIC PRESS, INC.
111 Fifth Avenue, New York, New York 10003

United Kingdom Edition published by
ACADEMIC PRESS, INC. (LONDON) LTD.
Berkeley Square House, London W1X 6BA

LIBRARY OF CONGRESS CATALOG CARD NUMBER: 70-117635

PRINTED IN THE UNITED STATES OF AMERICA

Contents

Preface

The ultimate goal of theoretical chemistry is the attainment of so thorough an understanding of atoms that their complete behavior under all conditions of chemical interest can be predicted, together with the physical and chemical properties of all substances and their mixtures. In other words, the cause-and-effect relationship between the nature of atoms and the nature of their combinations should become perfectly revealed.

One of the central problems in the pursuit of this unattainable yet irresistibly challenging goal has been to understand the nature of chemical bonds. To understand bonds, one must be able to calculate their energies. This book tells how. It reports the first generally successful calculation of more than 850 kinds of bonds in more than 500 compounds. Such calculations provide fascinating new insights regarding the nature of bonds. In turn, these insights permit the first successful explanations of many previously puzzling phenomena, which are also included. I find these ideas immensely helpful in the classroom, and hope my fellow teachers will share this experience. The work should be equally useful to students and practicing chemists.

The advent of quantum mechanics stirred high hopes that the whole of chemical science could be created from fundamental theory. An abysmal gap soon became evident, however, between principle and practice. Quantum mechanics has been of indispensable assistance in the development of modern chemical theories of atomic structure, and atomic and molecular spectroscopy, and in many other areas. But even the heaviest artillery of quantum mechanics, brought to fullest effect over a period of forty years through the medium of modern computers, has scarcely dented the problem of bond energy calculation.

There are two good reasons for this failure. One is the immense complexity of the practically insoluble problem of calculating all the interactions among all the component particles of an atom. The other is the fact that the energies of interactions among atoms are usually far smaller than the total energies of the atoms. The logical calculation of bonding energy as a difference between the total energy of a molecule and the total energy of its atoms is therefore subject to the difficulty of obtaining accurately very small differences between very large values. Compare, for example, the energies of two oxygen atoms (obtained as the sum of the successive ionization energies), roughly 94,000 kcal per mole, with the O_2 bond energy of 119. Even very good approximate solutions of the many-body problem could hardly be expected to provide reliable bond energies.

The approach I have taken, therefore, over the past twenty years, has been to accept the findings of quantum mechanics to the limit of their usefulness, and then deliberately to avoid the insoluble many-particle problem by attempting to identify those qualities of an atom which in a sense summarize, or are the resultant of, all its interelectronic and electron-nuclear interactions. I have now identified these qualities as the covalent radius, the electronegativity, and the homonuclear bond energy, and have shown how the latter two are interrelated and can be obtained one from the other. These atomic properties, plus the bond length, are the basic data for bond energy calculations as described herein.

This work has revealed many questions needing answer, which I hope many readers will become interested in pursuing.

For financial support, I am indebted to the University of Iowa and especially to Arizona State University for having provided me with a steady salary and comfortable working conditions. For their moral support I am grateful to my wife Bernice and my son Bob, and to my respected colleagues Dr. LeRoy Eyring, Dr. Sheng Lin, and Dr. Paul Stutsman, who have sympathetically strengthened my philosophical endurance of the frustrations of frequent opposition.

None of this work would have been possible without the help of contributions from both experimental and theoretical chemists far too numerous to acknowledge individually, but nonetheless deeply appreciated. Their data and ideas have been a constant source of inspiration.

R. T. SANDERSON

Chemical Bonds
and Bond Energy

Chapter 1

FUNDAMENTAL PRINCIPLES
AND SIGNIFICANT PROPERTIES

The structure of atoms and the nature of their bonds are of the most fundamental importance in chemistry. The extranuclear structure of atoms has become reasonably well understood. The nature of their bonds to other atoms remains, despite many years of prodigious efforts and remarkable achievements by skilled theoreticians and experimenters, a fertile field for further investigation. This book describes the results of one continuing study within this area which has developed a simple yet quantitative theory of bonding. Because of its quantitative nature this theory has proven uniquely useful in explaining chemistry. As a practical basis for understanding and because of its relative simplicity, it should have wide appeal to chemists in general.

Introduction to Bond Energy Calculation

If the net attractive interaction between two atoms or among more than two atoms is sufficiently strong so that the unique properties of the combination can be studied experimentally before it decomposes, the atoms are said to be held together by chemical bonds. Energetically, the combination corresponds to a system of minimum potential energy achieved when the atoms are brought close together from infinite separation. A bond is considered to exist between each pair of atoms in the combination which are in actual close contact with one another. The distance between the adjacent atomic nuclei corresponding to the potential energy minimum is called the bond length. Since at ordinary temperatures, combined atoms are vibrating back and forth along their bond axis as well as in other directions, the bond length refers to the atoms at their average, or equilibrium position with

1

respect to the internuclear distance. The potential energy, or energy required to return these atoms to an infinite separation, is called the **bond energy.**

To achieve an understanding of the nature of a chemical bond, it is necessary to be able to predict the bond length and the energy. To date, this has been accomplished by strict application of pure wave mechanics only for the H_2^+ molecule-ion, a system involving only one electron and two protons.[1] The energy of the hydrogen molecule, H_2, has been calculated accurately by James and Coolidge,[2] and later by others,[3] but only by assuming the experimentally determined bond length. In addition, a number of other small molecules have been subject to similar wave-mechanical calculations using modern high-speed computers with considerable success.[4] However, at best the methods are exceedingly complex, and usually cannot be applied without simplifying and somewhat arbitrary assumptions.

In principle, wave mechanics is widely accepted as providing the means for calculating all the properties of matter. *In practice*, it is recognized as being mathematically too laborious and difficult. Considering only the coulombic interactions among the component particles, one recognizes in the simple hydrogen molecule 6 such interactions. The two protons repel one another, the two electrons repel one another, and each electron interacts with each of the two protons. In the Li_2 molecule of only eight basic particles, 25 such interactions must similarly be taken into account. The problem becomes rapidly more complex with each increase in the number of fundamental particles in the molecule. We must therefore seek some less sophisticated methods which, while simple and explanatory, do not violate the basic truths or valid precepts of wave mechanics.

Such a search has been the purpose of the work described in this book. Most of the difficulty mentioned above can be avoided if the complex interactions within each atom, to the extent that they influence its interactions with other atoms, can be expressed as their resultant. In other words, one may recognize certain numerical quantities as fundamental atomic properties consistently shown by the atoms under all conditions of chemical interest. Specifically, it has now been recognized that the most fundamentally useful atomic properties are the **nonpolar covalent radius,** the **homonuclear bond energy,** and the **electronegativity.**

The primary purpose of this book is to show what can be learned about

[1] O. Burrau, *Det. Kgl. Danske Vid. Selskab.* **7,** 1 (1927); M. Born and J. R. Oppenheimer, *Ann. Phys.* (*Leipzig*) **84,** 457 (1927); others include E. A. Hylleraas, *Z.Phys.* **71,** 739 (1931); G. Jaffe, *ibid.,* **87,** 535 (1934); E. Teller and H. O. Sahlin *in* "Physical Chemistry" (H. Eyring, D. Henderson, and W. Jost, eds.), Vol. 5, p. 35. Academic Press, New York, 1970.

[2] H. M. James and A. S. Coolidge, *J. Chem. Phys.* **1,** 825 (1933).

[3] W. Kolos and L. Wolniewicz, *J. Chem. Phys.* **49,** 404 (1968).

[4] G. Das and A. C. Wahl, *J. Chem. Phys.* **44,** 87 (1966); G. Das, *ibid.,* **46,** 1568 (1967); and more recent papers by them and others, including E. Clementi.

chemical bonds from these simplified atomic properties through application of easily visualizable and rational concepts of bond formation. In particular, from a knowledge of these values plus the bond length, one may now calculate accurately the bond energies and therefore heats of formation of hundreds of both molecular and nonmolecular compounds. From these, the enthalpies of thousands of reactions can be calculated, and, more important, understood. The methods have proven to have valuable diagnostic power in aiding to identify the nature of a particular bond. The calculations have proved invaluable in explaining many heretofore puzzling chemical phenomena. They have also provided a new and superior model of nonmolecular bonding in binary salts. And, equally important, the results have revealed some of the questions we must ask in seeking deeper understanding.

One of these questions is, how is the strength of a bond influenced by the nature of the other bonds formed by the same atom at the same time? This is not a problem presented by diatomic molecules but it becomes very important for triatomic and more complex molecules. One must recognize here two hypothetical modes of synthesis of the molecule or nonmolecular solid. One method would be to bring the atoms together from infinite separation, *one by one*, letting each bond form before the arrival of the next atom. The other method would be to bring the atoms together from infinite separation *simultaneously*, so that the bonds all form at once when the atoms come close enough together. Reversal of these processes would require the input of what is called bond energy. The stepwise process would be exactly equivalent to the simultaneous process of atomization in that the sum of the steps must equal the whole. However, the energy of each individual step is not the same as the average for all the steps, even if the several bonds become ultimately identical.

For example, the total atomization energy of methane, CH_4, may be regarded as the sum of successive energies for the stepwise atomization, or as four times the average bond energy. However, the bond dissociation energy for the process

$$CH_4 \rightarrow CH_3 + H$$

is not the same as for the process

$$CH_3 \rightarrow CH_2 + H$$

nor is it equal to the average bond energy. We must therefore distinguish carefully between **bond dissociation energy** and **average bond energy.** The former is an important quantity for many purposes, but is *not* the direct concern of this book. We shall be interested only in total atomization energies or in individual bond energies such that their sum equals the atomization energy. The reasonable assumption consistently made is that if several bonds of the same kind are within the same molecule, they will contribute equally to its total energy, despite the experimental fact that if the bonds are broken stepwise, the energy requirements of the successive steps are different.

It will also be consistently assumed that if the sum of the individual calculated bond energies equals the experimental heat of atomization, the individual bond energies are also probably correct. The experimental basis for determining atomization energy is the thermodynamic one. The standard heat of formation of a compound is subtracted from the sum of the experimental atomization energies of the separate elements from their standard states. For many compounds and many of the elements these quantities have been determined very accurately. Some uncertainty exists in other reported values, which limits the confidence one may place in some of the calculations to be discussed. Unfortunately, where a discrepancy exists between the calculated and the experimental value, it will not always be possible to know whether this reflects an imperfection of the method of calculation or an experimental error. Fortunately, however, the number of reliable data is great enough and the extent of agreement between calculated and experimental values is sufficient to justify an overall confidence, tempered by reasonable discretion in certain instances.

General Method of Bond Energy Calculation

The general method of bond energy calculation, to be described in detail in Chapter 2, is based on the concept of a dual bond nature. A heteronuclear chemical bond is considered to combine the qualities associated with nonpolar covalence with those associated with electrostatic or ionic bonding. For purposes of simplification of the mathematics only, and emphatically not as literal description, the bond is treated *as if* it were nonpolar covalent part of the time, and completely ionic the rest of the time. The nonpolar covalent bond energy is calculated as the geometric mean of the homonuclear single covalent bond energies of the individual elements, but corrected for any difference between the covalent radius sum and the observed bond length. The ionic bond energy is calculated, for molecular compounds, as simply the coulombic energy of opposite charges at the observed internuclear distance.

For nonmolecular compounds, such as solid binary salts, the covalent energy is calculated as for molecular compounds except that it is multiplied by a factor determined by the coordination of atoms around a given atom. The ionic energy is calculated as the crystal lattice energy according to the Born-Mayer equation.

The bond energy in the molecular compound or the total atomization energy of the nonmolecular compound is then calculated as the weighted sum of the covalent and ionic energies. Highly critical is the assignment of weighting coefficients. These are based, as will be described, on the partial charges on the bonded atoms. These charges are calculated from the atomic electronegativities.

The atomic properties required for bond energy calculation according to this general method are therefore the covalent radius, the electronegativity, and/or the homonuclear single covalent bond energy (which as will be shown are closely related and can be derived one from the other). In addition, the experimental bond length must be known or a reasonable estimate must be possible. Since these properties are so fundamental and vital to all that follows, they will now be discussed in detail.

Fundamental Atomic Properties

Covalent Radius

The extent to which the electronic cloud about an atomic nucleus spreads out away from the nucleus must represent a balance between the effective nuclear charge and the repulsions among the several electrons. This balance does not lead, however, to a finite limit to the extension of the cloud, or a "skin" enclosing the atom at a fixed radius. Therefore an atomic radius can only be defined in terms of the approach of two atoms. Although the extent of the electronic cloud cannot be measured exactly, the positions of two nuclei can be detected experimentally in such a way that the internuclear distance can be accurately determined. The covalent radius of an atom can be defined as one-half the internuclear distance between it and a like atom when the two are joined by a single covalent bond and bear no electrical charge.

Accurate bond lengths for the homonuclear diatomic molecules of the alkali metals and the halogens are experimentally available. In addition, the bond lengths in the diamond structure of carbon, silicon, germanium, and tin are accurately known. For other elements, many of which do not form homonuclear single bonds, more indirect methods of estimating covalent radii have been employed. Although the literature presents reasonably good agreement among the several reported estimates for any given element, the data are not altogether self-consistent or beyond question. The following method has therefore been employed to evaluate the covalent radii of major group elements not reliably obtainable from experimental evidence.

Some years ago when the method of evaluating electronegativity from the relative compactness of atomic clouds was developed, a linear relationship was observed between the average electronic density of an atom, $Z/4.19r^3$ (where $4.19 = 4/3$ of π), and the number of outermost shell electrons within a period. Based on this relationship, the function Z/r^3 was plotted against number of outer shell electrons for the elements of M1, M4, and M7 of each period. Exact linearity was observed, except for period 3. Therein, for reasons not yet understood, it is the square root of the Z/r^3 function that is linear from sodium through silicon to chlorine. These results are shown in Fig. 1-1. The nonpolar covalent radii of the other elements were then

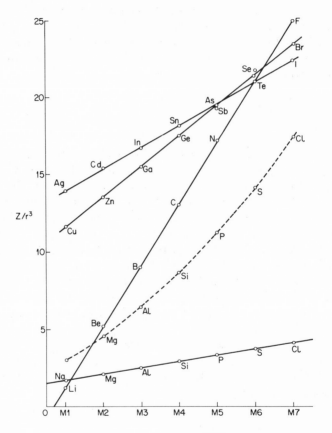

FIG. 1-1. Representation of $Z/r^3 = aM + b$. $\sqrt{Z/r^3} = aM + b$ for period 3.

determined by interpolation, except for zinc, gallium, cadmium, and indium for which the radii were extrapolated. Values obtained in this manner were essentially the same as those estimated by a variety of other methods. The method thus seems capable of providing an internally consistent set of covalent radii to be applied where needed in studying other properties of the atoms and their molecules. Results are listed in Table 1-1.

TABLE 1-1

Nonpolar Covalent Radii—Calculated and Experimental

	Radius (Å)		Experimental evidence
	Calc.	Exp.	
Li	1.34	1.34	R_o in $Li_2(g)$ 2.67
Be	0.91		
B	0.82		(B—B(g) 1.59)
C	0.77	0.77	R_o in diamond 1.54
N	0.74		
O	0.72		(O—F in OF_2 1.42)
F	0.71	0.71	R in $F_2(g)$ 1.42
Na	1.54	1.54	R_o in $Na_2(g)$ 3.08
Mg	1.38		
Al	1.26		
Si	1.17	1.17	R_o in Si(diamond) 2.35
P	1.10		(Black P 2.17–2.20)
S	1.04		R_o in S_8 (g) 2.08, (c) 2.10, polysulfides 2.05 av.
Cl	0.99	0.99	R_o in $Cl_2(g)$ 1.988
K		1.96	R_o in $K_2(g)$ 3.92
Ca		1.74	
Zn	1.30		(Tetrahedral r 1.31)
Ga	1.26		(Tetrahedral r 1.26)
Ge	1.22	1.22	R_o in Ge(diamond) 2.45
As	1.19		(Tetrahedral 1.18, "normal" 1.21, As_4 2.43)
Se	1.16	1.17	(Tetrahedral 1.14, crystal 2.32, "normal" 1.17)
Br	1.14	1.14	R_o in $Br_2(g)$ 2.284
Rb	2.16	2.16	R_o in $Rb_2(g)$ 4.32
Sr		1.91	
Cd	1.46		(Tetrahedral 1.48)
In	1.43		(Tetrahedral 1.44)
Sn	1.40	1.40	R_o in gray Sn, 2.80
Sb	1.38		(Tetrahedral 1.36, "normal" 1.41)
Te	1.35		(Tetrahedral 1.32, "normal" 1.37)
I	1.33	1.33	R_o in $I_2(g)$ 2.66
Cs	2.35	2.35	R_o in $Cs_2(g)$ 4.70
Ba		1.98	
Cu	1.35	1.35	(Tetrahedral 1.35)
Ag	1.50		(Tetrahedral 1.53)

Period	a	b
2	3.99	− 2.75
3	0.41	1.32
4	2.00	9.63
5	1.42	12.51

$r = \sqrt[3]{Z/(aM+b)^2}$ for period 3

$r = \sqrt[3]{Z/(aM+b)}$ for period 2 and 18-shell periods 4 and 5

ELECTRONEGATIVITY

Many different methods have been applied to the evaluation of the relative electronegativity of the chemical elements with surprisingly good agreement despite the lack of uniform definition of the property. There is little to be gained here by reviewing these many methods or attempting a critical evaluation of them because all are largely empirical. Qualitatively, all are alike in showing consistently upward trends across each period from alkali metal to halogen. Furthermore, there is general agreement that a bond between two atoms with different electronegativities will be polar to the extent of the difference, the initially more electronegative atom gaining negative charge at the expense of the other. However, there is one conspicuous difference between one method and all the rest. The one method referred to is that of basing electronegativity values on the relative compactness of the electronic clouds of the different atoms.[5] The logic of this method lies in the qualitative concept that if an atom holds its own electronic cloud tightly, despite the interelectronic repulsions, it must be expected to attract an outside electron strongly if an orbital vacancy exists to accommodate it. On the other hand, if an atom cannot even hold its own electronic cloud tightly against the interelectronic repulsions, it can hardly be expected to attract an outside electron strongly even though ample orbital vacancies may exist. An average electronic density was defined as $3Z/4\pi r^3$, where Z is the atomic number and $4/3\pi r^3$ is the volume of the electronic sphere defined by the nonpolar covalent radius r. The average number of electrons per cubic angstrom thus calculated was found to correspond in a general way to the accepted electronegativity values of the elements.

To permit a comparison between average electronic densities of the active elements and those of Group M8, the inert elements, radii of the latter were estimated by extrapolation of the radii of isoelectronic series of cations and anions. These estimated radii are now believed to correspond to hypothetical "ions" of zero charge, significantly different from "covalent radii" that would apply to M8 compounds. They were used for determining average electronic densities of the M8 atoms, which were found to vary somewhat despite the assumption that electronegativity was (at that time) meaningless in these elements. The average electronic densities of the active elements were then corrected for the changes in electronic density unrelated to electronegativity by taking as a measure of electronegativity the ratio of the actual average electronic density to that of a hypothetical inert element of the same atomic number (determined by interpolation of the plot of atomic number versus M8 element electronic density).

[5] R. T. Sanderson, "Chemical Periodicity," p. 26. Reinhold, New York, 1960; R. T. Sanderson, "Inorganic Chemistry," p. 72. Reinhold, New York, 1967.

The conspicuous difference mentioned above between this electronegativity scale and the Pauling scale[6] (an arbitrary scale to which it has apparently been felt necessary to convert all values from other methods) is that the others are linear functions of the Pauling electronegativity but the relative compactness scale values are linear with the *square root* of the Pauling values. As will be detailed later, the success of the relative compactness values in permitting useful estimations of the partial charges on combined atoms has seemed to eliminate any need of converting to the admittedly arbitrary Pauling scale.

Of all the various methods of evaluating electronegativity, the one having greatest intuitive appeal is that of Allred and Rochow.[7] They defined the electronegativity as the coulombic force interacting between an outside electron at the covalent radius from the nucleus, and the nuclear charge as shielded by the intervening electrons. Slater[8] had defined the "effective nuclear charge" as the difference between the actual nuclear charge and the sum of the screening constants of the electrons in the cloud. He assigned roughly reasonable values to these screening constants according to the type of energy level the electron occupied. Allred and Rochow used Slater's screening constants to compute the effective nuclear charge Z^*. They set electronegativity as proportional to the function Z^*e^2/r^2, where r is the covalent radius. Their electronegativities were converted to the Pauling scale by a linear equation. Since this work was completed, Clementi and Raimondi[9] have made much more sophisticated wave-mechanical calculations of the screening constants. When these are used in the Allred-Rochow method, minor changes are observed but the relative scale remains essentially the same.

Let us then adopt the relative compactness electronegativities, as they have been corrected or modified recently on the basis of considerations to be outlined presently. But let us also accept the idea that electronegativity must be a force between the nuclear charge and an electron at the periphery (covalent radius) of the atom. For this idea leads us toward an extremely interesting correlation, and perhaps eventually will justify a new definition of electronegativity based quantitatively on experimentally measurable properties.

HOMONUCLEAR SINGLE COVALENT BOND ENERGY

Another atomic property that might well be useful in predicting the

[6] L. Pauling, "Nature of the Chemical Bond," 3rd Ed., Cornell University Press, Ithaca, New York, 1960.

[7] A. L. Allred and E. G. Rochow, *J. Inorg. Nucl. Chem.* **5**, 264, 269 (1958)

[8] J. C. Slater, *Phys. Rev.* **36**, 57 (1930).

[9] E. Clementi and D. L. Raimondi, *J. Chem. Phys.* **38**, 2686 (1963).

behavior of atoms toward other atoms is the homonuclear single covalent bond energy. This is the energy of a single covalent bond between like atoms that are uncharged. Pauling[10] suggested the reasonable assumption that an atom should contribute to a nonpolar covalent bond with a different atom the same amount that it contributes to a similar bond to a like atom. We shall find no reason to doubt the validity of this assumption, which therefore becomes very useful. Experimentally, such homonuclear single covalent bond energies are available as dissociation energies of diatomic molecules of the alkali metals and the halogens, and from other thermo-chemical data for various other elements. In Chapter 4, the complete details of obtaining homonuclear single covalent bond energy parameters for the active elements are discussed. For the present, let us turn our attention to at least the qualitative aspects of the origin of such energy.

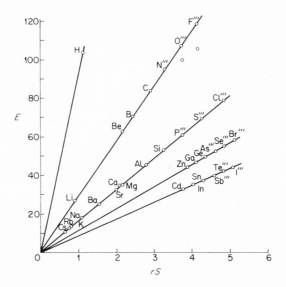

Fig. 1-2. Homonuclear single covalent bond energy and radius—electronegativity product.

Ultimately, it would seem that all net attraction between atoms must be of electrostatic origin. Specifically, the homonuclear single covalent bond energy must reflect the extent to which the positive charges on the two like nuclei, shielded as they must be by the presence of intervening electrons, interact with the two electrons that form the bond. This picture presents an extremely complex problem if it must be dealt with quantitatively in terms of all the interelectronic and internuclear and electronic–nuclear interactions

[10] L. Pauling, footnote 6, pp. 79, 82.

that must exist. A simple qualitative concept is that here too the effective nuclear charge must be considered. However, we are now talking about bond energy, not force. The homonuclear single covalent bond energy may be regarded as proportional to the coulombic energy of the attractions between the shielded nuclei and the bonding pair of electrons. That is, the energy E is proportional to $-Z^*e^2/r$.

If S (the coulomb force as electronegativity) is proportional to $-Z^*e^2/r^2$, and E (the homonuclear single covalent bond energy) is proportional to $-Z^*e^2/r$, then E should also be proportional to rS.

To test this very simple hypothesis, the known values of E were plotted against rS as shown in Fig. 1-2. A series of straight lines converging at the origin resulted. Each line consisted of all those elements having a certain easily definable electronic type. Their slopes were found to be directly related to the electronic type. That is, all the lines could be expressed by the linear equations $E = CrS$, where C is a proportionality constant. All the values of C, in turn, could be related to an integer n which represents the electronic type of atomic structure. The results are summarized in Table 1-2.

TABLE 1-2

RELATIONSHIP BETWEEN ELECTRONEGATIVITY, HOMONUCLEAR BOND ENERGY, AND ELECTRONIC CONFIGURATION: $E = CrS^a$

Equations: $C = \dfrac{37.0}{n-0.70} = E = \dfrac{37.0rS}{n-0.70}$ $S = \dfrac{E(n-0.70)}{37.0r}$

n	Electronic type of atom	C
1	(Hydrogen)	123.3
2	2 Electrons in penultimate shell	28.5
3	8 Electrons in penultimate shell	16.1
4	Period 4: 18 electrons in penultimate shell	11.2
5	Period 5: 18 electrons in penultimate shell	8.6

^{a} This equation was first empirically derived [R. T. Sanderson, *J. Inorg. Nucl. Chem.* **28,** 1553 (1966)].

It must be pointed out here that the homonuclear single bond energies for the elements from groups M5[11] through M7 in Fig. 1-2 are not the same as those usually accepted from thermochemical evidence. As will be explained in Chapter 3, in all those elements having lone pairs of electrons in the outermost shell there is evidence that such pairs are somehow capable of seriously weakening the bond. The values indicated in the figure are those

[11] Groups of the periodic table are numbered herein according to the suggestion [R. T. Sanderson, *J. Chem. Educ.* **41,** 187 (1964)] that major groups be denoted by M and transitional groups by *T*.

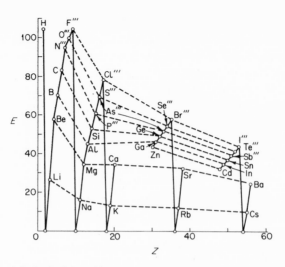

FIG. 1-3. Periodicity of homonuclear single covalent bond energy.

that would apply if all such bond weakening were removed. Ways of removing this weakening do exist and will be discussed in detail in later chapters.

For the present, the most important application of this work results from the **relationship between electronegativity and homonuclear single covalent bond energy.** One can easily be estimated or computed from the other.

The periodicity of homonuclear covalent bond energy is depicted in Fig. 1-3.

Chapter 2

METHODS OF CALCULATING
HETERONUCLEAR BOND ENERGY

Single Bond Energies in Molecular Compounds

The energy of a heteronuclear bond can be treated *as though* the bond were part nonpolar covalent and part ionic. The actual bond is of course believed to be a blend of these qualities. There is no intent to imply that it somehow alternates between the two. However, the mathematical treatment is greatly simplified by separating these two contributions and dealing with them individually. This division may well seem quite arbitrary, but in fact, it works remarkably well. One should keep in mind the fact that an arbitrary model may be found to be quantitatively consistent with experimental evidence and still fail to represent reality accurately. It is only fair to maintain, therefore, a degree of skepticism until the evidence favoring the validity of the model becomes completely convincing. At present, all that can be said for the model of the chemical bond proposed here is that it provides internal self-consistency, a logical visualizability, and extraordinarily successful results in the calculation of bond energies and heats of formation.

Nonpolar Covalent Bond Energy

In his early work in the development of the concept of electronegativity, Pauling[1] suggested that if a heteronuclear bond were not polar, its energy should be the average of the homonuclear bond energies of the two atoms involved. For practical reasons he made greater use of the arithmetic mean, but he expressed some preference for the geometric mean. Following Pauling, let us assume that each atom does contribute to the covalent part of its bond energy in proportion to its homonuclear single covalent bond energy,

[1] L. Pauling, "Nature of the Chemical Bond," 3rd Ed. Cornell University Press, Ithaca, New York, 1960.

no matter to what other atom it may become attached. We may then write for the nonpolar covalent bond energy, E_c, the simple expression:

$$E_c = \sqrt{E_A E_B} \tag{2-1}$$

E_A and E_B are the homonuclear single covalent bond energy parameters of elements A and B.

This expression, however, takes no account of bond length. It would be a valid expression if the actual bond length were exactly the sum of the nonpolar covalent radii of atoms A and B. Experimental measurements have shown abundant examples of bonds whose lengths do not obey this additivity rule, however. In general, the presence of bond polarity is associated with a shortening of the bond. In some compounds, the bond is longer than the covalent sum. For these reasons, plus the fact that the fundamentally electrostatic attraction of bond energy varies inversely with the distance between the opposite charges, expression (2-1) should be corrected for deviations of the bond length from the nonpolar covalent radius sum.

A simple correction factor that has proven effective is the ratio, R_c/R_o, in which R_c is the nonpolar covalent radius sum and R_o is the observed bond length. The nonpolar covalent bond energy can then better be expressed as:

$$E_c = \frac{R_c}{R_o} \sqrt{E_A E_B} \tag{2-2}$$

IONIC BOND ENERGY

The ionic energy of a bond is that which results from unequal sharing of the bonding electrons, which imparts a partial negative charge to one of the atoms and a partial positive charge to the other. The energy for the separated ionic component E_i of the total energy is therefore simply a coulomb energy:

$$E_i = 332e^2/R_o \tag{2-3}$$

The electronic charge e is taken as unity, and the factor 332 is that required to convert the energy to kilocalories per mole (kpm).

TOTAL BOND ENERGY

Pauling assumed that the ionic energy of a bond is a supplement to the normal nonpolar covalent energy. That is, he took as the ionic energy the difference between the experimental energy and the nonpolar covalent energy. **The present method makes the important distinction that ionic energy does not** *supplement* **covalent energy, but rather,** *substitutes for part* **of the covalent energy.** The separate contributions to the total energy are therefore weighted. The covalent weighting coefficient t_c is related to the ionic weighting coefficient t_i by the equation $t_c + t_i = 1.00$. The expression for the total bond energy then becomes:

$$E = t_c E_c + t_i E_i \qquad (2\text{-}4)$$

Bond Energies in Nonmolecular Binary Solids

COVALENT ENERGY

The covalent energy contribution to the total atomization energy of a nonmolecular solid is calculable exactly as for a bond in a molecular compound. The only difference is that the expression for covalent energy must now be multiplied by n, a factor indicating the degree of coordination around the central atom. The physical model implied by this treatment will be described and discussed in detail in Chapter 9. The covalent energy is then:

$$E_c = \frac{nR_c \sqrt{E_{AA} E_{BB}}}{R_0} \qquad (2\text{-}5)$$

IONIC ENERGY

The ionic energy of the nonmolecular solid is calculable as the crystal lattice energy according to the Born-Mayer equation:

$$E_i = \frac{332 M k \, z^+ z^- \, e^2}{R_0} \qquad (2\text{-}6)$$

In this expression, M is the Madelung constant, k the repulsion coefficient, z^+ and z^- the charges on the ions, and 332 again the factor for conversion to kilocalories per mole.

TOTAL ATOMIZATION ENERGY

The expression for the total atomization energy of a nonmolecular solid is the same as given above in Eq. (2-4).

We now turn attention to the all important question of how to evaluate the weighting coefficients.

Electronegativity Equalization

The term "electronegativity" as used in chemistry means "tendency to acquire extra electrons or to *become* negatively charged," and not, "quality of being negatively charged." Unfortunately, conventional usage is often at fault in implying that the electronegativity of an atom is an invariant quality independent of the condition of the atom with respect to chemical combination. For example, authors may write of the "highly electronegative fluorine in sodium fluoride." It seems completely logical, on the other hand, to recognize the limited capacity of atoms for extra electrons, and that this capacity must diminish to the extent that it is filled. Thus, although fluorine is indeed the most highly electronegative of all the chemical elements, the fluoride ion can have no electronegativity at all. It seems reasonable to suppose that in intermediate states of partial ionicity, fluorine must have intermediate electronegativity.

The recognition of this feature of the concept of electronegativity was published in 1951[2] as the "principle of electronegativity equalization." This principle may be stated as follows: **When two or more atoms initially different in electronegativity combine chemically, they become adjusted toward an equal intermediate electronegativity in the compound.** A simple physical picture of this equalization process is possible. For example, let us take a diatomic molecule of a binary compound, such as KBr. The potassium atom is initially very low in electronegativity, in keeping with its electronic structure in which the outermost shell electron is well shielded from the nuclear charge by the intervening electrons. The bromine atom is initially quite high in electronegativity. This is consistent with its position near the end of a period, throughout which the effective nuclear charge has been increasing by about two-thirds of a proton with each increase in atomic number, corresponding to the addition of electrons to the same outermost shell. Each atom has one half-filled outermost orbital, imparting the ability to form one covalent bond. The two atoms combine through a mutual sharing of the two bonding electrons, but the electrons tend to spend more than half time more closely associated with the bromine nucleus than with the potassium nucleus.

The effect of this uneven sharing is of course opposite for each atom. To the bromine atom it imparts partial negative charge. On the average the bromine electronic cloud now has a population greater than the original thirty-five electrons. This excess of negative charge increases the interelectronic repulsions causing the cloud to expand. With this expansion comes reduced opportunity of the electrons to penetrate toward the nucleus, the result of more complete shielding of the nuclear charge as well as greater average electron–nuclear distance. Consequently the electronegativity of

[2] R. T. Sanderson, *Science* **114,** 670 (1951). See also: R. Ferreira, *Trans Faraday Soc.* **59,** 1064 (1963); J. Hinze, M. A. Whitehead, and H. H. Jaffe, *J. Am. Chem. Soc.* **85,** 148 (1963); G. Klopman, *J. Am. Chem. Soc.* **86,** 1463, 4550 (1964); N. C. Baird, J. M. Sichel, and M. A. Whitehead, *Theoret. chim. Acta.* **11,** 38 (1968); M. C. Day and J. Selbin, "Theoretical Inorganic Chemistry," 2nd Ed., 138, Reinhold, New York, 1969; "Physical Chemistry" (H. Eyring, D. Henderson, and W. Jost, eds.) Vol. 5, pp. 182, 268, Academic Press, New York, 1970; G. Van Hooydonk and Z. Beckhaut, *Chem. Ber.* **74,** 323, 327 (1970); R. S. Evans and J. E. Huheey, *J. Inorg. Nucl. Chem.* **32,** 373, 383, 777 (1970).

the bromine atom is diminished. Simultaneously the uneven sharing results in a deficiency of electrons in the potassium cloud, imparting a partial positive charge. The significance of this partial charge is that it implies a less well shielded nucleus, a higher effective nuclear charge, and a smaller, more tightly held electronic cloud. Closer approach of the bonding electrons to the potassium nucleus is thus permitted during that fraction of the time when they are more closely associated with the potassium. In other words, the electronegativity of the potassium increases.

It is reasonable to suppose that this adjustment in electronegativities of the two atoms will cease at the point where equalization is reached. The bonding electrons thus become equally attracted to both nuclei by virtue of being unevenly shared. By "equally attracted" is meant the following. During the smaller fraction of time around the potassium, the bonding electrons have essentially the same energy as they have during the greater fraction of time around the bromine. In other words, equalization of electronegativity is equivalent to equalization of orbital energy. Equalization would seem a necessary corequisite to the merging of atomic orbitals to form molecular orbitals.

The next problem is to determine exactly what the intermediate electronegativity in the compound is. Here the choice of averaging technique seems somewhat arbitrary, but experience has shown that the geometric mean is very satisfactory. As a corollary to the principle of electronegativity equalization, then, it may be stated that the electronegativity reached in the compound is the geometric mean of the electronegativities of all the individual atoms before combination. If we know how much the electronegativity of each atom changes in the process of forming the compound, we can then correlate this change with the amount of partial charge that accompanies it. The calculation of partial charge is one of the most important applications of electronegativity that can be made.

Partial Charge

Although many attempts to estimate ionicity of bonds have been made, none has been free from objection. Perhaps the most widely recognized method is that of Pauling,[3] who found an empirical relation between the electronegativity difference and the ionicity as measured by the dipole moment. The dipole moment has more recently been recognized as being extremely sensitive to minor changes in the electronic clouds of atoms and therefore very difficult to relate directly to bond polarity. For example, the dipole moment of HCl is only 16% as great as it would be if unit positive

[3] L. Pauling, "Nature of the Chemical Bond," 3rd Ed. Cornell University Press, Ithaca, New York, 1960.

charge resided at the hydrogen nucleus and unit negative charge resided at the chlorine nucleus. This leads to the conclusion that the H—Cl bond is 16% ionic (which happens to be correct). However, the dipole moment is the product of two unknowns, the charge and the separation. Only by assuming one can the other be estimated from the dipole moment. The positive charges in HCl total 18, and the electronic charges also 18. The experimental dipole moment actually measures the net moment produced by a separation of a center of a total of $+18$ charge from the center of a total of -18 charge. The separation is only a very small fraction of the bond length. A shift of only a few hundredths of an angstrom unit in the center of negative charge would cause a relatively large change in the dipole moment. The dipole moment is therefore a very dubious indication of the exact bond polarity.

Even more objectionable is the use of electronegativity difference as though the bond were dependent only on the nature of the two atoms, ignoring any and all other atoms of the molecule. In an extreme example, this would imply that the bond polarity of the O—H bond must be the same in OH⁻ ion as in H_3O^+ ion. A much more reasonable assumption is that every bond in a molecule will be influenced by every other bond, an assumption inherent in the principle of electronegativity equalization. Therefore the ionicity curve of Pauling, even as later revised by Hannay and Smyth,[4] could at best be applicable only to diatomic molecules.

By far the simplest and most successful method of evaluating the ionicity of a chemical bond is that of calculating partial charge on the basis of the equalization of electronegativity. Two assumptions were originally made. One was that electronegativity changes linearly with partial charge acquisition. The other is the assumption of a specific ionicity of 75% in sodium fluoride. This assumption is required by the fact that no experimental evidence exists that unambiguously establishes the exact ionicity of any bond, except through subjective interpretation which may well be fallacious. The original assumption was 90%.[5] However, chemical intuition soon suggested a reduction to 75%.[6] At the time these assumptions were made their purpose was only to establish a relative scale of partial charges. It was hoped that they would be of the right general order of magnitude. It was not expected that they would have absolute validity, without further correction, for which no basis was then known.

The general scheme of calculating partial charges from these assumptions is thoroughly detailed elsewhere[7] and need only be outlined here. The

[4] N. B. Hannay and C. P. Smyth, *J. Amer. Chem. Soc.* **68,** 171 (1946).
[5] R. T. Sanderson, *J. Chem. Educ.* **31,** 2, 238 (1954).   [6] *Ibid.,* **32,** 140 (1955).
[7] R. T. Sanderson, "Chemical Periodicity," Reinhold, New York, 1960; *ibid.,* "Inorganic Chemistry," Reinhold, New York, 1967.

electronegativity of NaF is calculated as the geometric mean of the electro-negativities of sodium and fluorine. The differences between this value and the original electronegativities of sodium and of fluorine are measures of the change in electronegativity undergone by each atom in forming the molecule. Then if the electronegativity of fluorine, for example, was reduced by this difference while the fluorine acquired a partial charge of -0.75, the assumption of linearity of electronegativity change with charge would give the change that would correspond to -1.00 charge, as the observed change divided by 0.75. This change corresponding to acquisition of unit charge happens to be 4.99 for fluorine. The partial charge on any atom of combined fluorine is then defined as the ratio of the change in electro-negativity undergone in forming the compound, to 4.99. More generally, any partial charge on any atom is defined as **the ratio of the electronegativity change undergone in forming the compound to the change that would correspond to the acquisition of unit (+ or −) charge.** Thus, if the electronegativity of fluorine happened to change by 2.50 in a compound, the partial charge on fluorine would be $2.50/4.99 = 0.50$, to which a minus sign would be attached indicating a reduction in electronegativity of the fluorine.

In a similar manner, the change corresponding to formation of sodium ion was calculated. Starting with these two elements, further calculations then resulted in tabulation of unit charge electronegativity changes for all the elements for which electronegativities were known. A somewhat com-plicated algebraic calculation then disclosed that this procedure was equivalent mathematically to calculating the unit charge change as $2.08\sqrt{S}$:

$$\Delta S_{A\text{-}A}(+ \text{ or } -) = 2.08\sqrt{S_A} \qquad (2\text{-}7)$$

The factor 2.08 was determined by the selection of 75% for the ionicity of NaF. Any other selection would have established a valid *relative* scale but changed the factor.

A very simple general method of calculating relative partial charges was thus developed, applicable to practically any compound at all of known formula if composed of elements of known electronegativities. The procedure was first to determine the geometric mean electronegativity and then, for each atom, to calculate the ratio of the actual change in electronegativity to that expected for unit charge. The necessary data for performing such calculations are given in Table 4-1, page 57.

Weighting Coefficients

Because the choice of 75% ionicity for NaF was only based on chemical intuition and therefore expected only to be of approximately the right order of magnitude, the partial charges so obtained were considered to have relative validity only. When the need for applying partial charges to the determination of weighting coefficients for bond energy calculations became apparent, it was fully expected that a correction factor converting the relative to absolute charges would be required. Investigation soon demonstrated that no such correction was necessary. In other words, the choice of 75% ionicity was indeed a lucky one. As will be demonstrated later, in detail, application to sodium fluoride crystal gives an atomization energy of 182.4 kcal per mole, to be compared to the experimental value of 181.9. Indeed, bond energy calculations for hundreds of bonds have now been successfully accomplished. None of these would have been possible if the partial charge values were not correct. This point deserves special emphasis (see below) because of the apparent fact that partial charges, first developed and published in 1954, have not been widely accepted, despite their demonstrated wide applicability to chemical interpretations.[8]

For a bond in a molecular compound, the ionic weighting coefficient t_i, is simply the average of the partial charges on the two atoms forming the bond:

$$t_i = \frac{\delta_A - \delta_B}{2} \tag{2-8}$$

$$(t_c = 1.00 - t_i)$$

The implication for molecules having more than two atoms each is that the ionic form of the bond no longer involves unit charge on each atom. That is, when $t_i = 1.00$ as in the ionic form of a diatomic molecule, the partial charges in a molecule AB_2 would be 1.33 on A and -0.67 on B; in a molecule, AB_3, the partial charges would be 1.50 on A and -0.50 on B. However, in no known compound are such charge values even closely approached. The above formula for t_i is accurate for all the hundreds of molecular compounds studied thus far. For a nonmolecular compound, t_i is simply the partial charge on one atom divided by its oxidation number.

In order to demonstrate the validity of weighting coefficients derived from partial charge, and therefore the validity of partial charge itself, the data in Table 2-1 are presented. For this table, seventeen representative gaseous binary compounds were selected to cover a range of t_i values all the way from 0.03 to 0.81. For each compound, the total bond energy was calculated, assuming first, $t_i = 0$, and second, $t_i = 1.00$. As shown in columns 3 and 4 of Table 2-1, the ionic energy is much larger than the covalent energy, by a factor of 2–5. If it is reasonable to expect the true energy to lie between these extreme values, the weighting coefficient must be accurate or the chance of calculating the correct atomization energy is indeed small.

[8] Ref. 7; see also R. T. Sanderson; *J. Chem. Educ.* **41**, 331, 361, 415 (1964).

In columns 5 and 6 are given the calculated atomization energies based on the t_i values of column 2, and the experimental energies. The following questions seem reasonable: How could such remarkable agreement result if the concept of partial charge and its numerical evaluation were invalid? Indeed, what do these results seem to indicate about the validity of the entire procedure?

TABLE 2-1

COVALENT, IONIC, AND CALCULATED AND EXPERIMENTAL
ATOMIZATION ENERGIES OF SOME REPRESENTATIVE GAS MOLECULES

Compound	t_i	Total atomization energy (kpm)[a]			
		If covalent ($t_i = 0$)	If ionic ($t_i = 1.00$)	Calc.	Exp.
CH_4	0.03	370.8	1216	396.4	397.6
CS_2	0.04	263.6	640	278.8	276.4
ClF_3	0.09	96.8	404	124.3	124.7
NH_3	0.11	194.1	975	280.2	280.3
SO_2	0.12	209.7	580	254.2	256.7
HCl	0.16	73.6	261	103.7	103.3
H_2O	0.18	120.8	692	223.6	221.6
XeF_4	0.20	78.4	340	130.8	128
HF	0.25	60.9	360	135.9	135.8
$SiCl_4$	0.27	277.2	657	380.0	382.3
B_2O_3	0.28	405.4	1292	653.4	649.5
BF_3	0.36	292.5	765	462.9	463.0
AlF_3	0.48	243.6	611	420.0	421.2
KI	0.63	23.5	108.8	77.3	77.8
RbBr	0.73	26.7	112.5	89.4	90.4
KCl	0.76	30.8	124.2	101.9	101.6
CsCl	0.81	28.9	114.1	97.9	101

[a] kpm = kilocalories per mole.

The Physical Model of Bonding

POTASSIUM BROMIDE GAS MOLECULE

Gaseous potassium bromide can serve as an example of a bond energy calculation for a simple diatomic molecule. We will ignore the occasional assertion that KBr is completely ionic and exists as an ion pair in the gas phase. Instead we will explore the consequences of the initial electronegativity differences, consistent with the previous qualitative discussion.

The electronegativity of potassium is 0.42 and that of bromine, 4.53. Their geometric mean is 1.38. This means that in forming the KBr molecule, the electronegativity of potassium has increased by $1.38 - 0.42 = 0.96$. The complete removal of one electron from a potassium atom would have increased the electronegativity by 1.35. The partial charge on the potassium atom in KBr is therefore $0.96/1.35 = 0.71$. Similarly, the electronegativity of bromine has decreased by $4.53 - 1.38 = 3.15$. Had the bromine gained an electron completely, it would have lost 4.43 in electronegativity. Therefore the partial charge on bromine is $-3.15/4.43 = -0.71$. The bond is therefore 71% ionic. The ionic weighting coefficient is $t_i = 0.71$. The covalent weighting coefficient, t_c, is $1.00 - 0.71 = 0.29$.

The dissociation energy of the K_2 gas molecule is 13.2 kpm (kilocalories per mole). That of the Br_2 gas molecule is 46.1 kpm. The geometric mean of these is 24.6 kpm. The nonpolar covalent radius of potassium is 1.96 Å and that of bromine, 1.14 Å. The nonpolar covalent radius sum is then 3.10 Å. The observed bond length, 2.82 Å, shows the usual contraction for a polar covalent bond. We may now calculate the total bond energy in gaseous KBr:

$$E = \frac{0.29 \times 24.6 \times 3.10}{2.82} + \frac{0.71 \times 332}{2.82} = 7.8 + 83.6 = 91.4 \text{ kpm}$$

The atomization energy of potassium from its standard state is reported as 21.3 kpm. That of bromine is 26.7. The standard heat of formation of KBr(g) is -43.0 kpm. The sum of the atomization energies of the elements minus the heat of formation gives the total atomization energy of KBr: $21.3 + 26.7 - (-43.0) = 91.0$ kpm. The experimental value is thus in excellent agreement with that calculated above. To determine the calculated standard heat of formation, subtract the sum of the experimental heats of atomization of potassium and bromine from the calculated heat of atomization of KBr and change the sign: $91.4 - 21.3 - 26.7 = 43.4$; -43.4 kpm is the calculated standard heat of formation of KBr(g). The experimental value is -43.0 kpm.

The KBr molecule can therefore be regarded as a combination of atoms so initially different in electronegativity that the bond energy consists of a relatively large ionic contribution substituting for most of the covalent energy. However, the transfer of an electron from potassium to bromine is sufficiently incomplete to cause the molecule to behave, for purposes of calculation, as though it were nonpolar covalent 29% of the time. This contributes only 8.5% of the total bonding energy.

In general, since the largest known single covalent bond energy for a homonuclear bond is only 104 kpm (for H_2), and all other energies are substantially smaller, the factor 332 used in computing the ionic energy contribution gives an important advantage to the ionic energy by making it disproportionately large. For this reason, **bond strength is always enhanced by ionicity.**

POTASSIUM BROMIDE CRYSTAL

With the single exception of the boron halides, all known compounds in which the simplest molecule would still contain one or more outer shell vacant orbitals as well as outer shell lone pairs of electrons tend to combine further, giving more highly associated states. As will be detailed further in Chapter 9, molecules of metal combined with nonmetal are most likely to be of this type. The metal atom in the molecule will usually have vacant orbitals readily available and the nonmetal atom will have lone pairs available. Furthermore, the partial positive charge on the metal atom will enhance its electron pair acceptor ability. The partial negative charge on the nonmetal atom will enhance its electron pair donor ability. Continuing coordination is therefore the rule, to which a molecule of KBr is no exception. With partial charge of 0.71 on potassium and -0.71 on bromine, we must expect continuing association until each potassium atom is surrounded by bromine atoms and each bromine atom by potassium atoms. In this particular compound, the rocksalt structure is assumed, each atom being octahedrally surrounded by six atoms of the other kind. The Madelung constant is 1.75 and the repulsion coefficient is 0.91. The internuclear distance in the crystal is 3.30 Å. Since a total of eight electrons is available for six bonds, the equivalent number of covalent bonds, n, is 4. The covalent energy contribution is

$$t_c E_c = \frac{4 \times 24.6 \times 0.29 \times 3.10}{3.30} = 26.8 \text{ kpm}$$

The ionic contribution is

$$t_i E_i = \frac{0.71 \times 332 \times 1.75 \times 0.91}{3.30} = 113.8 \text{ kpm}$$

The sum, 140.6 kpm, is the total atomization energy calculated for KBr crystal. The experimental value is 142.1 kpm. Subtracting the atomization energy from the sum of the atomization energies of potassium (21.3) and bromine (26.7) gives -92.5 kpm, the calculated standard heat of formation. The experimental value is -94.1 kpm.

Observe that the covalent contribution is 19.1% of the total bonding energy in the crystal compared to only 8.5% in the gas molecule. This shows that the tendency toward further association from the gaseous state is encouraged not only by the increased electrostatic energy but also by the energy of coordination. The calculated energy of condensation is $-140.6 + 91.4 = -49.2$ kpm. The experimental value is $-142.2 + 91.0 = -51.2$ kpm. The methods of bond energy calculation have thus provided a means not only of predicting the condensation of KBr gas, but also of determining quantitatively the energy of the condensation.

Attention may now be called to what will be emphasized in Chapter 9. With only slight and reasonable modification, the method applicable to gaseous molecules is also applicable to solids of nonmolecular type. A relatively simple physical concept of the nature of polar bonding has sufficed to support an effective quantitative evaluation of chemical bond strength in both gas and solid.

The example of potassium bromide is only one of many that might have been given, and for which data are indeed presented later in this book.

Accuracy of Calculations

There is no possibility of assigning consistent limits of error in the calculation of bond energy, chiefly because no appropriate experimental standards are available. If it would be presumptuous of me to question the validity of some of the experimental bond energies, heats of formation, and bond lengths reported in the literature, then let the reported values speak for themselves. Lack of agreement is common, not only among different experimenters but even between different experiments reported at different times by the same experimenter. Differences among different individual measurements very commonly exceed significantly the probable limits of error set by each investigator. For many compounds, heats of formation and bond lengths have been measured very precisely by several methods independently by several investigators with excellent agreement. Values thus obtained can be accepted with maximum confidence. For many other compounds, and especially those having only fleeting existence or measurable only at very high temperatures, the data are inevitably less reliable, often less accurate, and occasionally very inaccurate as judged by large discrepancies among reported values.

Nevertheless the question of accuracy of bond energy calculations must not be evaded, for it is of central importance in this entire exposition. A complete analysis has been made of the percentage differences between experimental and calculated atomization energies of all the more than 500 compounds mentioned in this book. The results may be summarized as follows:

For nearly half (46.5%) of the compounds studied (which include practically all gaseous molecules for which experimental heats of formation could be found as well as a hundred binary solids) the calculated atomization energies agree within 1% of the experimental values. About 75% agree within 3%, and more than 84% of the compounds within 5%. The agreement is within 10% for about 93% of these compounds.

As will be further detailed in the following chapter, in every instance a reasonable, if not always predictable, bond type was assigned to give closest agreement. The significance of poor agreement, therefore, is that either the experimental value or the method of calculation is in error. Judgment as to which must depend on each individual example. The good agreement is so extensive and consistent that a reasonable case can be made for regarding disagreement as most commonly an indication of experimental error, if there is no obvious reason why the calculation methods should suddenly become invalid. Unquestionably, certain refinements of the methods will become possible when better and more complete experimental data become available. However, it may be possible now to draw some useful conclusions from an analysis of the kinds of bonds exhibiting wide discrepancy between calculated and experimental energies.

TABLE 2-2

COMPOUNDS FOR WHICH CALCULATED ATOMIZATION
ENERGIES DIFFER BY 5% OR MORE FROM THE EXPERIMENTAL VALUES

High-temperature gases			Molecular addition	Miscellaneous compounds	
LiH	BeCl	TlBr	BH_3CO	LiH(c)	$SbCl_3(g)$
LiF	BeBr	TlI	$BH_3N(CH_3)_3$	LiBr(c)	$SbBr_3(g)$
LiCl	BeI	CF	BH_3PF_3	NaH(c)	$OF_2(g)$
LiBr	CaO	CCl	$BH_3S(CH_3)_2$	K_2S(c)	$S_2Cl_2(g)$
LiI	BaF	Si_2	$BH_3S(C_2H_5)_2$	MgS(c)	$SF_5Cl(g)$
NaH	BN	$SiCl_2$	$(CH_3)_3BNH_3$	SrS(c)	$H_2Se(g)$
NaF	BS	GeF	$(CH_3)_3BNH(CH_3)_2$	BaS(c)	$SeBr_2(g)$
NaCl	BF	GeBr		ZnI_2(g)	BrF(g)
NaBr	BCl	$GeBr_2$		BH_3(g)	BrCl(g)
NaI	BBr	SnH		CCl_4(g)	ICl(g)
KH	BI	PbF		CCl_3F(g)	IBr(g)
KF	AlF	NS		C_2Cl_4(g)	$XeO_3(g)$
RbF	GaH	PF_2		SnO_2(c)	MnO(c)
CsH	InH	TeSe		PbS(c)	CuI(c)
CsF	TlF	ClO		AsF_3(g)	
BeH	TlCl				

Table 2-2 lists all compounds studied in which the calculated atomization energies differ from the experimental values by 5% or more. Of 87 compounds listed, 48 are gaseous molecules, mostly diatomic, that exist in the gas state only at high temperatures, where experimental bond energy measurements appear to be especially susceptible to error. Another 7 are molecular addition compounds, to be discussed in Chapter 8, and the remainder are mostly miscellaneous gaseous halides and solid chalcides. In summary, with the exceptions noted, no special class or group of compounds has yet been found to be consistently outside the scope of applicability of the calculation methods.

Chapter 3

BOND MULTIPLICITY
AND SUPER SINGLE BONDS

The methods described in the preceding chapter have been applied to a large number of both molecular and nonmolecular compounds with an extraordinarily high degree of success. For a considerable number of molecular compounds, however, the calculated bond energies or atomization energies based on ordinary single bonds are very appreciably lower than the experimental values. Two types of higher energy bonds are observed.

First, many of these molecules contain bonds that have long been recognized as involving more than two valence electrons per atom pair. Such bonds are called multiple bonds, or bonds of higher bond order than one. When four electrons per atom pair are involved, they are known as double bonds, and when six electrons are involved, triple bonds. Many bonds are also recognized as being of nonintegral bond order. Experience has shown that all such bonds characteristically have more than the normal single covalent bond energy and are generally shorter in length. Since bond multiplicity was not considered in the preceding chapter, it is appropriate here to study its effects and learn how to assess them quantitatively.

For the second type of higher energy bond, multiple bonding of the type mentioned above seems an impossible explanation. The most common examples are found in molecules containing oxygen or a halogen. These are compounds usually explained in terms of "pi bonding," involving partial "multiplicity" through supplementary donor–acceptor interaction. The recent bond energy work reveals a new viewpoint of this kind of bonding, which will also be examined in this chapter.

Multiplicity Factors

As will be discussed in detail in Chapter 10, the C—C and C—H bond energies are slightly variable but as a fairly close approximation may be considered constant, at least in hydrocarbons. Taking the C—C energy as 83 and the C—H energy as 99 kpm, one can then use the experimental heat of atomization of an olefin or acetylene to determine by difference the multiple bond energy. This procedure was used for a number of olefinic double bonds and a number of acetylenic triple bonds. A calculation was then made to determine by what factor the single bond energy must be multiplied to equal the multiple bond energy, if the single bond energy is first corrected to the double or triple bond length. The correction is necessary because even without multiplicity, a single covalent bond between carbon atoms would be stronger if shorter, and multiplicity always shortens the bond.

The results of these calculations are shown in Table 3-1 for double bonds

TABLE 3-1

DERIVATION OF DOUBLE BOND FACTOR

1.48	Ethene
1.52	Propene
1.51	1-Butene
1.50	1-Pentene
1.52	2-Pentene
1.54	2-Methyl-1-butene
1.52	3-Methyl-1-butene
1.56	2-Methyl-2-butene
1.50	1-Hexene
1.48	1-Heptene
1.49	1-Octene
1.47	1-Nonene
1.46	1-Decene
1.50	Average value

TABLE 3-2

DERIVATION OF TRIPLE BOND FACTOR

1.77	Ethyne (acetylene)
1.81	Propyne
1.77	1-Butyne
1.72	1-Pentyne
1.78	2-Pentyne
1.77	Average value

and in Table 3-2 for triple bonds. It can be seen that the multiplicity factor for a double bond is 1.50, and for a triple bond, 1.77. The latter is known with less certainty. A theoretical derivation of these factors would be extremely interesting and important. At present, such an explanation is not available. As empirical factors, however, they appear quite reliable and accurate. For example, the triple bond factor appears accurately applicable to the calculation of the bond energy of N_2, P_2, CO, CS, and SiS, as well as all carbon–carbon triple bonds in acetylenes and derivatives. The double bond factor is accurately applicable to the carbon–oxygen double bond in

TABLE 3-3

SOME APPLICATIONS OF THE DOUBLE BOND FACTOR 1.50

Molecule	Atomization energy (kpm)	
	Calc.	Exp.
CO_2	384.2	384.6
CS_2	278.8	276.4
CH_3CH_2CHO	936.5	935.3
O_2	119.2	119.2
S_2	102.5	102.4
NO	151.6	151.0
SN	123.7	116.6
C_2	143.5	142.3
Si_2	83.3	77

TABLE 3-4

SOME APPLICATIONS OF THE TRIPLE BOND FACTOR 1.77

Molecule	Atomization energy (kpm)	
	Calc.	Exp.
N_2	226.0	226.0
P_2	125.1	125.1
C_2H_2	391.0	392.6
CO	261.6	257.3
CS	167.8	166
SiS	154.1	158.6

all carbonyl compounds and CO_2 to O_2, S_2, NO, SN, C_2, Si_2, and other gaseous molecules. Tables 3-3 and 3-4 summarize these results.

The multiplicity of carbon–carbon bonds in graphite, benzene, and in intermediate aromatic structures must also be taken into account. Again empirically, the factor for all aromatic carbon–carbon ring bonds is found to be represented by $(1 + 0.33n)$, where n is the average fraction of a pi electron per bond. For example, in benzene each ring bond involves on the average 1.00 pi electron, so the multiplicity factor is 1.33. In graphite, each ring bond involves on the average 0.67 pi electron. Here the multiplicity factor is 1.22. For all intermediate condensed ring systems the multiplicity factors are between these two limits, and calculable from the simple expression $(1 + 0.33n)$. Table 3-5 summarizes some applications of such factors. Empirically it can be shown that for more than two electrons in a bond, the multiplicity factor varies approximately linearly with the cube root of the number of bond electrons.

TABLE 3-5

SOME APPLICATIONS OF MULTIPLICITY FACTORS TO AROMATIC RINGS

Molecule	No. pi electrons per ring bond	Factor	Atomization energy (kpm)	
			Calc.	Exp.
Benzene, C_6H_6	1.00	1.33	1319.4	1320.6
Phenol, C_6H_5OH	1.00	1.33	1421.1	1421.7
Naphthalene, $C_{10}H_8$	0.91	1.30	2094.5	2094
Anthracene, $C_{14}H_{10}$	0.87	1.29	2873	2865

A very important feature of multiple bonding, and one which distinguishes it from the "pi bonding" to be discussed below, is the applicability of the multiplicity factor to the *total* energy of the bond, not merely to the covalent contribution. For example, a carbon–oxygen bond is quite polar, in CO_2 giving an ionic weighting coefficient of $t_i = 0.17$. The ionic energy contribution is 38% of the total. Calculation of the atomization energy of carbon dioxide, multiplying the total single bond energy by the factor 1.50, gives 384.2 kpm. The experimental value is 384.6 kpm, in practically perfect agreement. In contrast, the kind of super-energy to be considered next applies only to the *covalent* contribution to the total bond energy.

Super Single Bond Energies

The nitrogen–nitrogen single bond energy has long been known to be about 39.2 kpm. The nonpolar covalent radius of nitrogen is 0.74 Å, making the nonpolar single bond length 1.48 Å. In the nitrogen molecule, N_2, the bond is generally accepted as being a triple bond and the bond length is 1.10 Å. The dissociation energy of N_2 is 226.0 kpm. It should be apparent that even when the 39.2 value is corrected for the shorter bond length, multiplication by the triple bond factor 1.77 will not give anything near the experimental value of 226. In fact, $39.2 \times 1.77 \times 1.48/1.10 = 93.2$ kcal, only 41% of the experimental value. Then what is wrong with the multiplicity factor, or the calculation?

It is enlightening to calculate what the single bond energy would be if it were consistent with the observed triple bond energy. By performing the following calculation: $(226.0 \times 1.10)/(1.77 \times 1.48) = 94.8$ kpm, we learn that the nitrogen single bond energy parameter would have to be 94.8 instead of 39.2 to give the experimental dissociation energy of N_2. The new value is exactly that calculated from the $E = CrS$ equation of Chapter 1. It is the value of 39.2 that appears anomalously low. A little reflection on the bond energies of other elements will show that this is not the only example of an anomalously low single bond energy. Similarly low values have long been recognized for oxygen and fluorine.

Since the trend from lithium through carbon is uniform, it takes only a little thought to recognize that what is most conspicuously different about nitrogen, oxygen, and fluorine, is the presence of outermost shell lone pairs of electrons. Let us make the tentative assumption that it is one or more of these lone pairs that cause the anomalously low bond energies. We are not at this point fully prepared to speculate on the cause of this assumed lone pair effect. However, it may be supposed that an outer pair of electrons unoccupied in bonding might interfere with any bonds formed by the atom by somehow blocking off the interaction between the nucleus and the valence electrons. It may further be supposed that in the event of formation of a triple bond, as in N_2, the concentration of six bonding electrons within the internuclear region would tend to force the lone pairs to the far side of the atoms where they could perhaps not interfere at all. Let it be assumed, then, that the "unweakened" single bond energy of nitrogen is represented by the 94.8 value and the fully weakened single bond energy is represented by the 39.2 value.

In the light of these assumptions and this viewpoint, consideration of the situation in oxygen may prove useful. The dissociation energy of O_2 is 119.2 kpm. In contrast, the single covalent bond energy as determined from thermal data on hydrogen peroxide (Chapter 4) is only 34.0 kpm. The correction for shortened bond length in the double bond is certainly not sufficient to give a single bond energy which when multiplied by the olefinic factor 1.50 would come anywhere close to 119. Let us determine what the single covalent bond energy should be to correspond to the experimental double bond energy. Following the same procedure as for nitrogen, the single bond energy that should correspond to the observed double bond energy is $(119.2 \times 1.21)/(1.50 \times 1.44) = 66.7$ kpm. In this calculation, the bond length in O_2 is 1.21, the covalent radius sum 1.44, and the multiplicity factor for the double bond, 1.50.

The objection might be raised that the double bond in oxygen is not the same as a double bond in an olefin, and therefore the factor 1.50 should not be applicable. The molecular orbital interpretation of O_2 does indeed show that the "double" quality or order arises from the presence of six bonding electrons and two antibonding electrons. This is not the same as the situation in the $C=C$ double bond, where there are only four bonding electrons and no antibonding electrons. Nevertheless, there seems to be no basis for making a distinction as far as bond energy calculations are concerned. Oxygen forms many double bonds other than to itself, in which the factor 1.50 is accurately applicable as is the bond energy of 66.7 calculated by its use.

It is significant that the bond energy in O_2 is about 25 kpm lower than that in a carbon–carbon double bond (about 119 compared to 144). This fact suggests that the alleged weakening caused by the lone pair electrons of oxygen has not been completely removed when the single bond energy is raised from 34 to 66.7 kpm. The weakening effect may be assumed to be half removed in the formation of a double bond. Then removal of the weakening effect completely would correspond to a single bond energy of 99.4 kpm. Since oxygen forms an acetylenic triple bond in only one known molecule, CO, the opportunity for full testing of this hypothesis is very limited. However, the bond energy of CO is worth examining.

The partial charges in CO are carbon 0.16 and oxygen -0.16, giving a value of 0.16 for t_i and 0.84 for t_c. The covalent radius sum is 1.49 Å whereas the observed bond length is 1.13 Å. The geometric mean of the carbon energy 83.2 and the newly calculated oxygen energy of 99.4 is 91.0. The multiplicity factor, since the bond appears to be triple, is 1.77. The covalent energy contribution is then $(0.84 \times 1.77 \times 91.0 \times 1.49)/1.13 = 178.4$ kpm. The ionic energy contribution is $(0.16 \times 332 \times 1.77)/1.13 = 83.2$ kpm. The sum, 261.6, agrees reasonably well with the experimental dissociation energy of 257.3 kpm. The calculated standard heat of formation is -30.7, to be compared with the experimental value of -26.4 kpm. Although no other triple bonds to oxygen are known, a number of compounds will later be discussed in which the 99.4 bond energy value is applicable to single bonds. Therefore this value seems at least tentatively acceptable, as does the assumption that the value for the single bond energy corresponding to a double bond is halfway between the values that correspond to completely weakened and completely unweakened condition of the single bond.

These are the initial evidences from which, as well as from many other related observations, it is concluded that all the elements of the periodic major groups M5, M6, and M7 exhibit the phenomenon of "lone pair weakening" of their single covalent bond energy parameters. Let us accept, at least tentatively, three different bond energy parameters for each of these elements. These are distinguished from one another by single, double, and triple primes. For example, the N''' energy is 94.8, the N'' energy is 67.0, and the N' energy is 39.2 kpm. The O''' energy is 99.4, the O'' energy 66.7, and the O' energy 34.0 kpm. The triple prime energies in general are those determined by extrapolation of the $E = CrS$ lines of Fig. 1-2 in Chapter 1. These values can in some instances be confirmed empirically from experimental bond energies, as done above for N_2. The single prime bond energies are largely obtainable from experiment. Details of the evaluation of these different single bond energies for each element are discussed in the following chapter. The values are summarized in Table 3-6.

TABLE 3-6

BOND ENERGIES FOR M5, M6, AND M7 SINGLE COVALENT BONDS

	N′	39.2	O′	34.0		F′		37.8
	N″	67.0	O″	66.7		F″		69.1
	N‴	94.8	O‴	99.4 (107.0)[b]	F‴		100.3 (118.3)
W[a]		0.41		0.34 (0.32)			0.37 (0.32)
	P′	51.1	S′	55.0		Cl′		58.2
	P″	55.9	S″	62.1		Cl″		68.4
	P‴	60.7	S‴	69.0		Cl‴		78.6
W		0.84		0.80				0.74
	As′	43.4	Se′	44		Br′		46.1
	As″	47.7	Se″	49.3		Br″		52.0
	As‴	52.0	Se‴	54.6		Br‴		58.0
W		0.83		0.81				0.79
	Sb′	32.0	Te′	34		I′		36.1
	Sb″	35.8	Te″	38		I″		40.0
	Sb‴	39.6	Te‴	41.7		I‴		43.9
W		0.81		0.81				0.82

[a] $W = E'/E'''$.
[b] Value in parentheses = CrS values.

It is of interest that oxygen and fluorine still remain somewhat exceptional Their triple prime energies as determined by extrapolation of the $E = CrS$ line beyond nitrogen are appreciably higher than the values determined from experimental data on bond energies of oxides and fluorides. One must assume that for these elements only, some bond weakening influence in addition to that of the lone pair effect must exist. Even this influence is removed in the single example of the Si—F bond, of which there are several examples (Table 3-8) suggesting that the correct single bond energy for this application is the one obtained by linear extrapolation. The relationships of these different single bond energies are shown for the period 2 elements in Fig. 3-1, and for period 3 elements in Fig. 3-2.

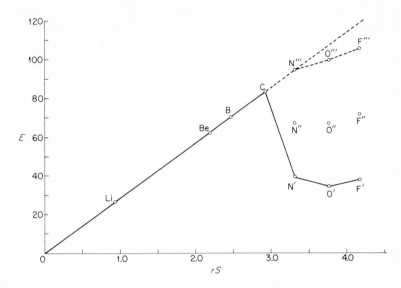

FIG. 3.1. Period 2 bond energy and radius — electronegativity product.

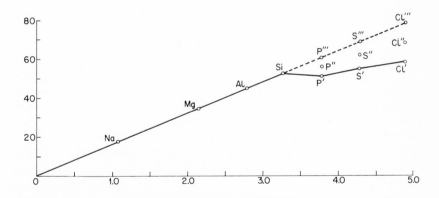

FIG. 3.2. Period 3 bond energy and radius — electronegativity product.

OCCURRENCE OF REDUCED WEAKENING

Before the problem of explaining the phenomenon of reduced weakening can even be approached, it is necessary to recognize the special circumstances that prevail where it occurs. These may be enumerated as follows:

1. Reduced weakening always accompanies bond multiplicity. The triple prime energy is always appropriate for a triple bond. A double prime (or

occasionally a triple prime) energy is always appropriate for a double bond.

2. When multiplicity is involved in resonance, between or among single and multiple bonds, the double or triple prime energy is appropriate for each of the bonds averaged in the molecule. For example, the two oxygen atoms that are bonded only to the nitrogen atom in HNO_3 are bonded equally, yet only one double bond is possible. The bond energies are correctly calculated as a single and a double bond energy but using the E'' energies for both.

3. When only a single bond is formed, the originally less electronegative atom is always one capable of supplying vacant orbitals to the originally more electronegative atom, which might accommodate one or more of its lone pairs. For example, the nitrogen–fluorine bond is an N—F′ bond but the boron–fluorine bond is a B—F‴ bond.

4. If the originally less electronegative atom lacks vacant orbitals of reasonably low energy, then the single bond only involves the E' energy, as just illustrated by N—F.

5. The carbon–fluorine bond, and possibly the carbon–oxygen bond on occasion, present the only known exceptions to rule 4. This is not exceptional if when the carbon has been subjected to substantial removal of its electrons, the orbitals of its $n = 3$ shell then become available through the increased effectiveness of the nuclear charge.

6. The occurrence of reduced weakening is recognizable with least ambiguity in compounds of nitrogen, oxygen, and fluorine where the differences between E' and E''' energies are much larger than for any of the heavier nonmetal atoms.

7. The O‴ energy is applicable only where oxygen is bonded to just one other atom. In other words, where oxygen forms a coordinate covalent bond by acting as electron pair acceptor, the single bond is —O‴. It appears to occur in POF_3, $POCl_3$, SOF_2, SO_2F_2, $SOCl_2$, and for two of the oxygen atoms in H_2SO_4. The occurrence of single bonds to oxygen is summarized in Table 3-7. Single bonds to oxygen employing the triple prime energy also appear to occur in resonance with double bonds to oxygen in gaseous SiO_2, SO, SO_2, and SeO_2.

8. The O″ energy seems applicable in some molecules where oxygen is bonded to just one other atom, as in the two single bonds that are in resonance with a double bond in SO_3, $POBr_3$, and SO_2Cl_2. It also applies to bridging oxygen, as in As_4O_6, solid SiO_2, $(HOBO)_3$, and $(BOF)_3$.

9. The extrapolated F‴ energy of 118.3 seems applicable only to the Si—F bond, as shown in Table 3-8. It appears that the silicon atom with its outer d orbitals activated through electron withdrawal by the fluorine must be unique in providing complete removal of the weakening effect of all kinds on the single bond energy of the fluorine.

TABLE 3-7

OCCURRENCE OF O'' AND O''' SINGLE BONDS

O″ single bonds			O‴ single bonds		
Compound	Atomization energy		Compound	Atomization energy	
	Calc.	Exp.		Calc.	Exp.
NO_2[a]	226.7	224.2	$SiO_2(g)$[a]	301.1	304.3
HNO_3[a]	373.5	376.2	POF_3	489.7	485.6
$POBr_3$	310.3	312.6	$POCl_3$	366.5	360.2
As_4O_6	934.8	935.8	SO[a]	122.0	124.7
SiO_2 (c)	441.2	445.8	SO_2[a]	254.3	256.7
SO_3[a]	341.6	340.0	$H_2SO_4(2)$	582.6	586.2
SO_2Cl_2	329.6	331.0	SOF_2	274.6	277.0
$(HOBO)_3$	1454.1	1463.2	$SOCl_2$	238.4	235.2
$(BOF)_3$	1205.1	1205.8	PO	125.3	125

[a] In resonance with double bond.

TABLE 3-8

APPLICATIONS OF HIGHEST FLUORINE SINGLE BOND ENERGY

Compound	δ_F	t_i	E_c	E_i	Atomization energy (kpm)	
					Calc.	Exp.
SiF	− 0.34	0.34	61.7	70.6	132.3	132
SiF_2	− 0.24	0.36	64.3	80.2	289.0	289
SiF_3	− 0.18	0.37	61.2	79.8	423.0	433.1
SiF_4	− 0.15	0.38	60.2	81.9	568.4	570.5
SiH_2F_2	− 0.33	0.35	61.6	73.5	444.0	444.9
$SiHF_3$	− 0.24	0.36	61.4	76.6	499.7	500.7
SiH_3F''[a]	− 0.40	0.33	48.2	68.9	386.2	389.1

[a] An apparent exception.

10. The F''' energy of 100.3 is applicable to the compounds listed in Table 3-9.

11. The F'' energy is applicable to the compounds listed in Table 3-10, and the F' energy to those in Table 3-11. It may be significant that the average value of t_i for compounds of Tables 3-8 and 3-9 is 0.37, for the metal salts of Table 3-10 it is 0.69, and for salts of Table 3-11 it is 0.82. Reduction in the lone pair weakening is evidently not favored by higher negative charge on fluorine or higher positive charge on the atom providing the vacant orbitals.

12. Similar results are found for chlorides, as listed in Tables 3-12 and 3-13, except that few if any applications of the Cl'' energy are known. Bromides and iodides are also similar, although reduction of the weakening appears somewhat less common in them.

The phenomenon of reduced weakening appears to occur only in molecular compounds and not in solid metal salts, for which the single prime energies are always observed. The higher single bond energy seems also to apply principally to the more electronegative atom of the molecule. Very few data are available on gaseous metal sulfides, selenides, arsenides, antimonides, etc., for comparison with the oxides and halides. With the limited and somewhat scattered data at hand, it is difficult to assemble an orderly, self-consistent picture from which some definite conclusions might be drawn. Certainly this is an area needing much further study. Some very interesting suggestions, however, can be distilled from the above collection of data and limited generalizations.

First, it must be observed that the phenomenon of abnormally weak single covalent bonds in nitrogen, oxygen, and fluorine has certainly not escaped attention until now. Indeed, many years of study, speculation, and suggestion have been devoted to this problem, especially that of the low dissociation energy of fluorine. The concept most generally accepted currently seems to be that independently suggested by both Mulliken[1] and Pitzer.[2] These men proposed, essentially, that the lone pairs on adjacent atoms, especially in the period 2 elements, may be so situated that formation of a single covalent bond between them results in appreciable repulsions between lone pairs on the adjacent atoms. In the light of the newer information derived from this study of bond energies, such repulsions no longer appear adequate to provide a satisfactory explanation. For in bonds to atoms such as hydrogen and carbon, which have no lone pairs in the outer shell, the same weak single bond energies apply as those that occur in the homonuclear single bonds. For instance, the bond energy of HF is very accurately calculable using the F' energy of 37.8 kpm. The calculated energy of 135.9 is practically identical with the experimental value of 135.8 kpm.

The conclusion seems inescapable that the phenomenon originates *within each atom* instead of representing an interaction between two atoms. Hence the assumption that somehow a lone pair interferes with the strength of the effective nuclear charge in the bonding direction seems more reasonable. The above observations do suggest the very intriguing idea that the sole contribution made by what is generally termed "pi bonding" is that of reducing the lone pair weakening. The strengthening effect of "pi bonding"

[1] R. S. Mulliken, *J. Amer. Chem. Soc.* **72,** 4493 (1950); *ibid.,* **77,** 884 (1955).
[2] K. S. Pitzer, *J. Amer. Chem. Soc.* **70,** 2140 (1947).

TABLE 3-9

APPLICATIONS OF F''' SINGLE BOND ENERGY

Compound (gaseous)	δ_F	t_i	E_c	E_i	Atomization energy (kpm)	
					Calc.	Exp.
BeF	− 0.41	0.41	53.6	96.5	150.1	146.9
BeF$_2$	− 0.29	0.44	50.2	102.2	304.8	303.6
BF	− 0.33	0.33	65.4	84.3	149.7	198.9
BF$_2$	− 0.23	0.34	64.4	86.8	302.4	302.3
BF$_3$	− 0.18	0.36	62.4	91.9	462.9	463.0
AlF	− 0.43	0.43	45.7	86.5	132.2	158.2
AlF$_2$	− 0.31	0.47	42.8	95.1	275.8	272.8
AlF$_3$	− 0.24	0.48	42.2	97.8	420.0	421.2
PF	− 0.26	0.26	60.4	54.3	114.7	112.7
PF$_2$	− 0.18	0.27	61.1	57.8	237.8	222.6
PF$_3$	− 0.14	0.28	60.7	60.4	363.3	356.1
PF$_5$ (3)	− 0.09	0.28	59.5	59.2	547.5	555.7
AsF$_5$ (3)	− 0.08	0.24	56.5	44.7	462.2	462.4

TABLE 3-10

APPLICATIONS OF F'' SINGLE BOND ENERGY

Compound (gaseous)	δ_F	t_i	E_c	E_i	Atomization energy (kpm)	
					Calc.	Exp.
MgF	− 0.55	0.55	26.3	104.3	130.6	129.6
CaF	− 0.62	0.62	22.4	101.9	124.3	127.
CaF$_2$	− 0.47	0.70	17.7	115.0	265.4	268.3
SrF	− 0.66	0.66	20.1	104.3	124.4	127.
SrF$_2$	− 0.51	0.76	14.2	120.2	268.8	262.8
BaF	− 0.73	0.73	13.9	111.7	125.6	136.
BaF$_2$	− 0.56	0.84	8.2	128.5	273.4	273.
COF$_2$	− 0.14	0.23	65.9	58.3	421.8	420.4
CF$_4$[a]	− 0.09	0.23	65.4	57.8	458.8	467.9
CF$_2$	− 0.15	0.23	66.4	58.7	250.2	250.
CF$_3$[a]	− 0.11	0.23	65.4	57.8	352.6	344.5
CH$_2$F$_2$	− 0.28	0.21	65.2	51.3	411.8	420.1
CHF$_3$[a]	− 0.19	0.22	65.8	54.9	445.8	444.6
CCl$_2$F$_2$	− 0.15	0.22	65.8	54.9	379.8	382.3
CClF$_3$[a]	− 0.12	0.23	64.9	57.4	420.1	423.1
SiH$_3$F	− 0.40	0.33	48.2	68.9	386.2	389.1
TiF$_4$	− 0.28	0.70	10.5	129.1	558.4	558.8

[a] In resonance with C—F′ bonds (2 C—F″ bonds is maximum).

TABLE 3-11

APPLICATIONS OF F′ SINGLE BOND ENERGY

Compound (gaseous)	δ_F	Atomization energy (kpm)	
		Calc.	Exp.
LiF	− 0.74	169.7	136.8
NaF	− 0.75	142.5	114.8
KF	− 0.85	136.1	118.1
RbF	− 0.86	131.5	115.7
CsF	− 0.90	129.7	116.2
NF	− 0.14	68.9	73.3
NF_2	− 0.09	137.8	140.7
NF_3	− 0.07	206.7	200-2
N_2F_4	− 0.09	308.0	303.3
OF	− 0.05	46.0	46.1
OF_2	− 0.04	96.0	89.8
SF_4	− 0.07	327.2	327.4
SF_6	− 0.05	490.8	471.8
SeF_6	− 0.05	422.4	430.
TeF_6	− 0.07	459.6	474.
HF	− 0.25	135.9	135.8

TABLE 3-12

APPLICATIONS OF Cl‴ SINGLE BOND ENERGY

Compound (gaseous)	δ_{Cl}	t_i	Atomization energy (kpm)	
			Calc.	Exp.
BeCl	− 0.32	0.32	112.4	92.9
$BeCl_2$	− 0.23	0.34	226.8	222.8
BCl	− 0.24	0.24	102.6	120.4
BCl_2	− 0.17	0.25	211.0	213.4
BCl_3	− 0.13	0.26	316.2	318.3
$AlCl_3$	− 0.19	0.39	307.5	305.0
$GaCl_3$	− 0.10	0.21	254.1	260.5
$InCl_3$	− 0.14	0.27	(221.3)	222.0
SiCl	− 0.26	0.26	95.0	92.6
$SiCl_2$	− 0.18	0.27	191.8	206.7
$SiCl_4$	− 0.11	0.27	380.0	382.3
$GeCl_4$	− 0.07	0.16	322.4	324.9
$SnCl_4$	− 0.10	0.24	308.0	307.1
$BiCl_3$	− 0.11	0.22	201.0	200.0

TABLE 3-13

APPLICATIONS OF Cl′ SINGLE BOND ENERGY

Compound (gaseous)	δ_{Cl}	t_i	Atomization energy (kpm)	
			Calc.	Exp.
LiCl	− 0.65	0.65	122.6	114.3
NaCl	− 0.67	0.67	105.8	99.0
KCl	− 0.76	0.76	101.9	101.6
RbCl	− 0.78	0.78	99.5	100.5
CsCl	− 0.81	0.81	97.9	101.
MgCl	− 0.47	0.47	97.4	71 ± 16
$MgCl_2$	− 0.34	0.51	203.2	201.9
CCl_4	− 0.06	0.15	347.6	312.3
CCl	− 0.13	0.13	85.1	68.4

on a single bond would therefore be only the indirect one of making it possible for the single bond strength to be increased.

Further speculation at this point would perhaps not be very beneficial. The subject will receive additional attention in various specific applications throughout the rest of this book. A thorough understanding of the phenomenon must, however, await further investigation.

Chapter 4

ATOMIC PROPERTIES
OF THE CHEMICAL ELEMENTS

Although generally good agreement is found in applying the inter-relationships among atomic properties discussed in Chapter 1, for general use in bond energy calculations an internally consistent set of data is desirable. Because of minor uncertainties in experimental bond energies and empirical electronegativity values, a somewhat arbitrary choice of cornerstone data is required. In this work it was decided that the N_2 dissociation energy of 226.0 kpm could be accepted, along with the multiplicity factor 1.77 found for triple bonds, the covalent radius of 0.74 Å, and the relative compactness electronegativity value of 4.49. It was also assumed that the empirical equation, $E = CrS$, is quantitatively accurate, and that the relationship between C and n is exact. The value of C for nitrogen (and hence all elements where $n = 2$) is calculated to be 28.5. For the elements of period 3, and also the heavier elements having penultimate shells of eight electrons, a good average value of C is 16.1. By solution of simultaneous equations, the empirical constants a and b in the relation, $C = a/(n - b)$ are found to be 37.0 and 0.70. C is then found to be 11.2 for $n = 4$ and 8.6 for $n = 5$.

With these data to provide the cornerstone, the appropriate data for the rest of the chemical elements can be subjected to such minor adjustment as necessary to provide a self-consistent set of values. As will be demonstrated, hundreds of successful bond energy calculations support the essential validity of the choices mentioned above.

In the equation $E = CrS$, if one assumes that C and r are accurately known, there is still room for mutual adjustment of E and S, since only the ratio E/S is fixed. Judicious adjustments have been made where appropriate to provide the best agreement with experimental evidence. Since such adjusting is of necessity a somewhat subjective process, the complete details are given in the following section for critical examination by the reader.

HYDROGEN

The experimental dissociation energy of H_2 at 25° is well established as 104.2 kpm.[1] The electronegativity of 3.55 was originally arbitrarily assigned as seeming to be of the correct order of magnitude, intermediate between boron and carbon but nearer to carbon than to boron. No basis for seriously questioning this value has arisen over a period of 16 years. On the contrary, it seems to give very reasonable values of partial charges in hydrogen compounds, which in turn are applicable to successful bond energy calculations. The covalent radius of hydrogen is not clearly established, for it appears to differ from its value in H_2 when in other combinations, even those with essentially nonpolar bonds. The observed internuclear distance in H_2 is 0.74 Å, equivalent to a radius of 0.37 Å. In CH_4, on the other hand, the bond length is 1.09 Å. Since the bond is believed to be only about 3% polar, the covalent radius sum should equal the bond length. The radius of carbon in diamond is 0.77 Å, from which the hydrogen radius is only 0.32 Å. This is the radius used in determining the nonpolar covalent radius sum in hydrogen compounds.

An unexplained mystery is the fact that the 104.2 kpm value appears to be accurately applicable to bond energy calculations involving hydrogen compounds along with the radius of 0.32 Å. If the hydrogen homonuclear covalent bond energy were corrected for the shortening of the radius from 0.37 to 0.32, it would be 120.6 instead of 104.2 kpm. There is no indication, however, that the "corrected" value would lead to successful bond energy calculations.

Application of the equation, $E = CrS$, to hydrogen using the value $r = 0.32$ and $S = 3.55$ together with the experimental dissociation energy gives $C = 91.3$. On the other hand, from the equation $C = 37.0/(n - 0.70)$, where $n = 1$ for hydrogen, $C = 123.3$. If C is 91.3, then the latter equation

---

[1] Unless otherwise specified, the thermochemical data used in this book are taken from one of the following sources: "Selected Values of Chemical Thermodynamic Properties," Circular No. 500, U.S. Bureau of Standards (1949); *Nat. Bur. Stand. (U.S.) Tech. Note* No. 270–3 (1968); *Nat. Bur. Stand. (U.S.) Tech. Note* No. 270–4 (1969); Joint Army, Navy, and Air Force Thermochemical Tables, Dow Chemical Co., Midland, Michigan; T. L. Cottrell, "The Strengths of Chemical Bonds." Butterworths, London, 1958.

Bond lengths and other structural data were taken largely from the following: A. F. Wells, "Structural Inorganic Chemistry," 3rd Ed. Clarendon Press, Oxford, 1962; "Interatomic Distances," Special Publication No. 11, The Chemical Society, London, 1958; "Interatomic Distances Supplement," Special Publication No. 18, The Chemical Society, London, 1965.

would correspond to a value of $n = 0.69$. Clearly hydrogen does not fit very well with the other elements into the $E = CrS$ relationship.

LITHIUM

Lithium vapor is mostly monatomic, but a small fraction near the boiling point consists of Li_2 molecules. Their dissociation energy at 25° is experimentally determined to be 26.5 kpm and their bond length 2.68 Å, from which the nonpolar covalent radius is 1.34 Å.

For $n = 2$, C in the equation $E = CrS$ is 28.5. To fit this equation, using the experimental bond energy and covalent radius, the electronegativity S must be revised downward slightly from the earlier value of 0.74 to 0.69. Alternatively, if the early electronegativity is correct (which seems intuitively more acceptable since a slightly higher value than that of sodium 0.70 would be expected), the corresponding bond energy would be 28.3 kpm. As will be seen, the limited data on lithium compounds offer no basis for choice, since both sets of values give very satisfactory agreement between calculated and experimental bond energies. A compromise would be to accept 0.74 for the electronegativity and the experimental value of 26.5 for E. These are the values adopted, along with 1.34 Å for the covalent radius.

BERYLLIUM

No experimental source of a Be—Be bond energy is known. Some indirect means of measurement is therefore required. This energy can be back calculated from the known experimental bond energies of beryllium com pounds, taking the average value as a reasonable estimate. Unfortunately a certain arbitrariness is inherent in this method, for both electronegativity and bond energy may be somewhat mutually adjustable and still give fairly consistent results. The procedure also assumes that everything else about the bond energy calculation methods is completely correct. On this basis, when S is 2.39 and $r = 0.91$ Å, the Be—Be energy back-calculated from the gaseous dihalide molecule atomization energies that are experimentally available is BeF_2, 58.2; $BeCl_2$, 55.3; $BeBr_2$, 59.4; and BeI_2, 61.8. It should be pointed out that these values, which average 58.7, are all based on the triple prime energies of the halogens, determined as described under the separate halogen elements. To the extent that they are reliable, they tend to confirm or support the validity of the halogen energies.

On the other hand, the X′ energies of the halogens are experimentally and reliably determined. These are the energies used in calculations involving the solid halides. Using the experimental atomization energies of the solid halides as the basis for back-calculating the Be—Be energy, the following only moderately satisfactory results are obtained: BeF_2, 60.2; $BeCl_2$, 59.2;

$BeBr_2$, 53.5; and BeI_2, 56.5. The average is 57.4, not very different from the average for the gaseous halides.

For $n = 2$, the value of C is 28.5 in $E = CrS$, from which one can calculate $E = 62.2$ kpm. Although this value is a trifle higher than that obtained by the back-calculations, it is adopted for beryllium compounds, giving heteronuclear bond energy values agreeing fairly well with those experimentally determined.

BORON

From the standard heats of formation of gaseous monatomic boron (134.5 kpm) and gaseous diatomic boron, B_2 (198.5 kpm), one can calculate a B—B energy of about 70.5 kpm. There is no assurance that this bond represents fairly the more typical bonding condition in which boron forms three covalent bonds, but it appears to be very similar. The bond length is a little less than twice the accepted nonpolar covalent radius of 0.82 Å being 1.59 Å. The experimental bond energy of 70.5 kpm may be corrected to the "more normal" bond length by multiplying by the factor 1.59/1.64, giving a B—B energy of 68.4 kpm.

Revising the electronegativity of boron slightly upward from 2.84 to 2.93, and applying the nonpolar covalent radius of 0.82 Å to the $E = CrS$ equation where $C = 28.5$ for $n = 2$, give a calculated 68.4 kpm for the homonuclear covalent bond energy of boron. This is the value adopted. As will be demonstrated later, it is consistent with known experimental bond energies in boron compounds, according to the methods of calculation described herein.

CARBON

Small but doubtless significant difficulties, which will be discussed more fully under the general subject of organic compounds, require an unfortunate arbitrariness in the assignment of a C—C bond energy. The atomization of graphite is now accepted as requiring an energy of about 171.3 kpm. The standard heat of formation of diamond from graphite is 0.45 kpm, which makes the atomization energy of diamond 170.8 kpm. Since each interior atom of diamond is joined to four neighbors by single covalent bonds, the process of atomization must involve the equivalent of breaking two single C—C bonds (four half-bonds per atom). Half the atomization energy is 85.4 kpm, which should therefore be equal to the C—C bond energy. However, where C—C linkages are less involved, or not involved at all, the value seems nearer to 83 kpm. By back-calculation from the experimental bond energies of some of its simpler compounds, we find the carbon bond energy to be as follows: CO_2, 83.2; CO, 78.8; CS_2, 81.3; COS, 84.7; and CH_4, 84.0 kpm.

The electronegativity of carbon as determined by the relative compactness method is 3.79, a value which seems very reasonable and consistent with other properties of carbon. The bond length in diamond as well as in many organic compounds is 1.54 Å, leading to a nonpolar covalent radius of 0.77 Å.

In the $E = CrS$ equation where $n = 2$ and $C = 28.5$, these values of r and S correspond to 83.2 kpm for the carbon bond energy. This is the adopted value.

Further discussion of carbon bond energies will be found in Chapter 10 on applications to organic chemistry.

NITROGEN

The electronegativity of nitrogen by the relative compactness method is 4.49. The nonpolar covalent radius is 0.74 Å.

An N' bond energy can be derived from a thermochemical consideration of ammonia (NH_3) and hydrazine (N_2H_4). The atomization energy of ammonia is that of $3 H = 156.3 + N = 113.0$, minus the standard heat of formation of gaseous ammonia, -11.0, $= 280.3$ kpm. The average N—H bond energy is one-third of this value, or 93.4 kpm. The heat of atomization of N_2H_4 is 226.0 for 2 N, plus 208.4 for 4 H, minus the standard heat of formation of gaseous hydrazine, 22.8 kpm, $= 411.6$ kpm.

As will be explained later, one can justify the assumption that the N—H bonds in hydrazine have the same average energy as in ammonia. Subtraction of $4 \times 93.4 = 373.6$ from the 411.6 gives 38.0 kpm for the N' bond energy. This is not very different from the value of 38.4 determined by Pauling from older data.[2] Alternatively, assuming the experimental heat of formation of hydrazine to be correct, as well as the method of bond energy calculation, one can back-calculate the N' bond energy to be 39.2 kpm. This is the adopted value.

In the preceding chapter, it was shown that one can calculate an N''' bond energy from the data on the N_2 molecule equal to 94.8 kpm, identical to that obtained by extrapolation in Fig. 1-2. An N'' bond energy is then the value halfway between or 67.0 kpm.

OXYGEN

By the relative compactness method the electronegativity of oxygen is 5.21. The nonpolar covalent radius is 0.72 Å.

A value of the O' bond energy can be estimated from the data for water and hydrogen peroxide, assuming that the O—H bonds are of equal energy in each compound. The justification for this assumption will be discussed

[2] L. Pauling, "Nature of the Chemical Bond," 3rd Ed. Cornell University Press, Ithaca, New York, 1960.

in a later chapter. The heat of atomization of H_2O is the sum of the atomization energy of H_2 (104.2) and half the atomization energy of O_2 (59.6) minus the standard heat of formation of gaseous water (-57.8) which equals 221.6 kpm. This is the enthalpy of breaking two O—H bonds, which also occur in H_2O_2. The standard heat of formation of $H_2O_2(g)$ is -32.6 kpm. The atomization energy of H_2 (104.2) plus that of O_2 (119.2) minus -32.6 equals 256.0 kpm, the atomization energy of H_2O_2. The energy of the O—O bond by difference is $256.0 - 221.6 = 34.4$ kpm. This is not very different from the 33.2 value similarly estimated by Pauling, using older data.[2]

Alternatively, one can also back-calculate what the O—O energy should be, if one accepts the standard heat of formation of H_2O_2 and the bond energy calculation to be accurate. This calculation gives the value of 34.0 kpm for the O' homonuclear covalent bond energy, which is adopted for this work.

In the preceding chapter, the derivation of an O" single bond energy was described. Assuming this value, 66.7, to be halfway between the O' and the O''' values, one can estimate the O''' value as 99.4 kpm. Application of the $E = CrS$ equation, however, gives 107.0 for the O''' value. In all compounds where an O''' energy might appropriately be used, the value of 99.4 gives good agreement between calculated and experimental bond energies whereas the 107.0 seems appreciably too high. As mentioned in the preceding chapter, one may tentatively conclude that additional bond weakening exists here beyond that caused by the lone pairs. A similar conclusion is necessary for fluorine, as will be seen below. Otherwise, extrapolation of the E vs. rS lines of Fig. 1-2 gives reliable values for the triple prime energies.

FLUORINE

The bond length in F_2 is 1.42 Å corresponding to a nonpolar covalent radius of 0.71 Å. The electronegativity is 5.75 by the relative compactness method.

The dissociation energy of F_2 as selected by the National Bureau of Standards from a variety of experimental evidence is 37.8 kpm. This is the energy that corresponds to a maximum of lone pair bond weakening and would therefore be designated F' energy.

As with oxygen, there is a substantial discrepancy between the extrapolated ($E = CrS$) energy and that which appears applicable as an F''' energy. The extrapolated value is 118.3, but this value is only applicable in Si—F bonds as noted in the preceding chapter. Bond energies in gaseous fluorides are usually considerably greater than those calculated using the 37.8 energy value for fluorine whenever vacant orbitals exist on the other atom that might accommodate lone pairs on the fluorine, even partially. Back-calculation from experimental atomization of several gaseous fluorides gives the following F''' energies: BeF_2, 98.2; BF_3, 100.7; AlF_3, 102.0; average

100.3 kpm. This value is adopted as the F''' energy. Except for the Si—F bond as mentioned, this gives much better agreement between experimental and calculated bond energies for gaseous fluorides than the value of 118.3. The F'' energy taken as halfway between 37.8 and 100.3 is 69.1 kpm. As will be seen, this value seems applicable to calculation of the bond energies of gaseous difluorides except that of beryllium.

SODIUM

Like lithium, sodium vapor is principally monatomic but contains a small concentration of Na_2 molecules in the vapor, especially near to the boiling point. The dissociation energy of these diatomic molecules at 25° is known experimentally to be 18.0 kpm. This is taken as the homonuclear single covalent bond energy. From the electronegativity, 0.70, and the nonpolar covalent radius 1.54 Å (half the bond length in the Na_2 molecule), the value of C in the equation $E = CrS$ is found to be 16.7, whereas for $n = 3$, the value is 16.1. This difference would correspond to an increase of the electronegativity of sodium to 0.72, hardly enough to justify an adjustment.

Generally satisfactory agreement between calculated and experimental bond energies of sodium compounds is obtained by use of $E = 18.0$, $S = 0.70$, and $r = 1.54$ Å.

MAGNESIUM

The nonpolar covalent radius of magnesium is 1.38 Å and the electronegativity by the relative compactness method is 1.56; where $n = 3$, $E = 16.1\ rS = 34.6$ kpm.

Since no Mg—Mg bond energies are known from experiment, the above bond energy of 34.6 must be tested by calculation of the atomization energies of magnesium compounds. Here a difficulty is encountered, in that nearly all calculated values tend to be too high by an average of about 10 kpm. Although it is possible to adjust both E and S together, several attempts to find values giving better agreement with experimental atomization energies failed. This suggested the possibility of a consistent error in the determination of the experimental atomization energies, which are the difference between the sum of the atomization energies of the separate elements and the standard heat of formation of the compound. The only value in common is the atomization energy of magnesium. This is presumably reliably reported as about 35.5 kpm.

There is really little other justification than the bond energy calculations for suspecting this value, except that it is out of line in M2, being lower than for calcium. However, if the atomization energy of magnesium were really about 45.5 kpm instead of 35.5, then the following calculated and experimental atomization energies result: MgF(g), 130.6, 132.0; MgF_2(g), 262.2, 258.9; $MgCl_2$(g), 203.2, 204.3; MgBr(g), 84.6, 85; $MgBr_2$(g), 177.2, 173.4; MgF_2(c), 358.4, 352.0; $MgCl_2$(c), 256.7, 256.9; $MgBr_2$(c), 224.2, 222.9; MgI_2(c), 180.0, 180.0; MgO(c), 245.8, 248.9 kpm.

Tentatively, magnesium is accepted as conforming with the $E = CrS$ relationship if its radius is 1.38, its electronegativity 1.56, and its bond energy 34.6 kpm, provided the atomization energy of magnesium metal is indeed some 12 kpm higher than literature values indicate.

ALUMINUM

Taking the electronegativity of aluminum as 2.22 and the nonpolar covalent radius as 1.26 Å, then from $E = 16.1 \, rS$ for $n = 3$ is obtained the homonuclear bond energy parameter of 45.0 kpm.

Since no experimental bond energy for Al—Al is available, some indirect procedure for testing the above value must be used. The aluminum trihalide gas molecules are all assumed to involve the X''' energy for the halogens. On this basis, one can back-calculate the following homonuclear bond energy for aluminium: AlF_3, 44.1; $AlCl_3$, 45.8; $AlBr_3$, 49.9; and AlI_3, 42.6; average, 45.6 kpm. Although some other appropriate combination of bond energy and electronegativity might give acceptable results in bond energy calculations, the above values are tentatively accepted as being very satisfactory for the limited data available for aluminum compounds: $E = 45.0$, $r = 1.26$, $S = 2.22$.

SILICON

The bond length in solid silicon, of diamond structure, is 2.34 Å. Half of this value, 1.17 Å, is the nonpolar covalent radius. The electronegativity of silicon is revised slightly upward from the older relative compactness value to 2.84.

The atomization energy of solid silicon is reported to have the experimental value of 108.9 kpm. If as with diamond, half of this is taken as the single bond energy, the homonuclear single bond energy is 54.5. If, however, as also with diamond, extra bond energy comes from having a close assemblage of like atoms, a slightly lower energy might be more appropriate. From the above radius and electronegativity values and the relationship $E = 16.1 \, rS$ for $n = 3$, the value of E is 53.4 kpm—a reasonable reduction from the 54.5 comparable to that observed in carbon. The 53.4 value gives very satisfactory bond energy calculations for silicon compounds and thus appears to be acceptable and is adopted here.

PHOSPHORUS

The covalent radius of phosphorus is 1.10 Å, half the bond length in black phosphorus, which is nearly the same as in white phosphorus, P_4.

The electronegativity is revised slightly upward from the original relative compactness value of 3.34 to 3.43. The relationship, $E = 16.1 \, rS$ gives 60.7 for the phosphorus bond energy. This, however, must be the value for the P''' energy, since the lone pair on phosphorus must weaken the bond as in nitrogen, but as pointed out earlier, to a much smaller degree.

To check this, let us consider the gaseous molecule, P_2. The bond length is 1.89 Å, instead of the covalent radius sum of 2.20 Å, and its dissociation energy at 25° has been reported as 125.1 kpm. Just as for N_2, one can calculate what single bond energy would apply if there were no lone pair interference or weakening of the bond. The bond is assumed to be a triple bond as in N_2. The corresponding single bond energy is then $(125.1 \times 1.89)/(1.77 \times 2.20) = 60.7$ kpm, in agreement with the calculated value given above. This value is therefore adopted for P'''.

The P' energy can be estimated from data for $P_4(g)$, by subtracting the standard heat of formation of the gas, $+14.1$ kpm, from the total atomization energy of 4 moles of red phosphorus (P): $4 \times 79.4 = 318.6$, and dividing by 6, the number of bonds in the P_4 tetrahedron. This gives 50.8 kpm for the P—P bond at the observed bond length of 2.21 Å. Corrected to 2.20 Å this becomes 51.1 kpm. This is the P' bond energy.

Following the established procedure of assuming the P'' energy to be midway between the P' and the P''' values, we find for P'' the value 55.9 kpm. These values permit bond energy calculations for phosphorus compounds in reasonably good agreement with those determined experimentally. In most applications merely the P' value is needed.

SULFUR

The nonpolar covalent radius of sulfur is taken as 1.04 Å, from a variety of S—S bond lengths ranging between 2.04 and 2.10 Å, and by interpolation of the Z/r^3 function between silicon and chlorine. The relative compactness electronegativity is 4.12. From these two values and the expression, $E = 16.1 \, rS$, the bond energy is found to be 69.0 kpm. This would correspond to the S''' energy.

Initial attempts to assign a homonuclear covalent bond energy to sulfur based on the experimental atomization energy of $S_8(g)$ gave unsatisfactory results, due to apparently inconsistent thermochemical data in the literature. It was therefore decided to assume that the standard heat of formation of $H_2S(g)$ is completely accurate as reported from experiment. Back-calculation from the experimental atomization energy should give a correct S' energy. The value thus obtained is 55.0 kpm.

The experimentally determined dissociation energy of $S_2(g)$ at 25° is 102.5 kpm. The bond length is 1.89 Å, in comparison to the covalent radius sum of 2.08 Å. The S_2 molecule is paramagnetic like O_2, giving evidence of two unpaired electrons. It therefore seems reasonable to assume that a double bond is present. Using the olefinic double bond factor of 1.50 as in a similar calculation for O_2, one can calculate the S'' energy as $(102.5 \times 1.89)/(1.50 \times 2.08) = 62.1$ kpm. Following the usual procedure of considering the S'' energy to be halfway between the S' and the S''' energies, we find the S''' energy to be practically the same as calculated above, 69.2 kpm.

From the experimental data, the bond energy in S_8 is calculated to be 63.5 kpm of bonds. This agrees reasonably well with the S'' energy of 62.1. It explains why a suitable S' energy was not obtainable from the thermo-chemistry of S_8. It also suggests that there is a special reason for the stability of S_8 rings, as will be discussed in Chapter 5.

CHLORINE

The bond length in the Cl_2 molecule is known to be 1.99 Å from which the nonpolar covalent radius of chlorine is 0.99 Å. The relative compactness electronegativity is 4.93, a value about which no serious questions have arisen through many applications.

The experimental dissociation energy of Cl_2 at 25° is 58.2 kpm. This value is taken as the Cl' homonuclear single covalent bond energy. In a number of its gaseous compounds, however, chlorine appears to contribute more to the total atomization energy than is indicated by the Cl' energy. Each of these compounds represents a combination of chlorine atoms with another atom that can provide vacant outer orbitals for somehow reducing the bond weakening effect of the chlorine lone pairs of electrons. Back-calculation of the Cl—Cl bond energy from atomization energies of these gaseous chlorides, and assuming the bond energy calculation methods to be correct, gives the following results: $SiCl_4$, 80.8; $GeCl_4$, 80.3; $SnCl_4$, 77.9; BCl_3, 80.4; $AlCl_3$, 75.2; and $BeCl_2$, 72.6. The average is 77.9 kpm, the "experimental" Cl''' energy.

Application of the $E = CrS$ equation where C is 16.1 gives the value of 78.6 kpm for the Cl''' energy. This agrees very well with the experimental value, and is adopted. The Cl'' energy halfway between is then 68.4 kpm.

POTASSIUM

Like the other alkali metals, potassium in the vapor state is slightly associated to K_2 molecules. The bond length is 3.92 Å, from which the non-polar covalent radius is 1.96 Å. The electronegativity is adjusted slightly downward from an earlier value to 0.42. For $n = 3$, which has been found to apply to all elements having a penultimate shell of eight electrons whether in period 3 or not, C is 16.1. The $E = CrS$ equation then gives 13.2 as the bond energy. This is also the experimental dissociation energy of $K_2(g)$.

CALCIUM

The nonpolar covalent radius of calcium is somewhat uncertain but is taken as 1.74 Å which appears satisfactory. The electronegativity is 1.22.

No direct experimental evaluation of a Ca—Ca bond energy is known. Some indirect means must be sought. Only gas phase data for the fluorides are known for calcium, strontium, and barium. In these compounds the covalent energy is too small to serve reliably as a basis for back-calculating the metal–metal energy. Back-calculations from the experimental atomization energies of the solid halides are complicated by the fact that different crystalline structures occur. Only CaF_2 has the fluorite structure. Both the chloride and the bromide crystallize in a deformed rutile structure in which each metal atom has two nearest neighbors and four more appreciably farther distant. The iodide has the cadmium iodide layer structure. The bond energy parameters for calcium calculated from CaF_2 (36.8) and CaI_2 (35.7) average 36.3 kpm. Complications in the chloride and bromide calculations suggest that they be omitted from the present evaluation for later discussion.

Application of the $E = CrS$ equation gives a homonuclear single covalent bond energy of 34.2 kpm. Since this appears to be approximately correct, it is tentatively adopted.

ZINC

The nonpolar covalent radius of zinc extrapolated from bromine and germanium using the Z/r^3 function is 1.30 Å. Available thermochemical data for zinc compounds do not provide a very satisfactory basis for evaluating the homonuclear single covalent bond energy by back-calculation. However, results agree reasonably well with the value of 43.4 kpm, which is based on an electronegativity value revised somewhat upward from the earlier 2.84 to 2.98, and calculated from the $E = CrS$ equation where $C = 11.2$ corresponding to $n = 4$.

GALLIUM

The nonpolar covalent radius of gallium extrapolated from iodine and gray tin using the Z/r^3 function is 1.26 Å and the electronegativity is taken as 3.28. From the $E = CrS$ equation the homonuclear single covalent bond energy is calculated as 46.3 kpm. This is tentatively adopted.

Thermochemical data for gallium compounds are inadequate for reliable estimation of the bond energy by back-calculation. The dissociation energy of $Ga_2(g)$ is calculated to be 27.6 kpm from the enthalpies of formation of 66.2 for Ga and 104.8 for Ga_2. The lack of agreement is duly noted but as yet there is insufficient evidence for judgment.

GERMANIUM

The experimental bond length in germanium is 2.44 Å corresponding to a nonpolar covalent radius of 1.22 Å. The electronegativity is 3.59. The experimental atomization energy of germanium is 90.0 kpm. Since atomization of its diamond structure would involve the equivalent of breaking 2 Ge—Ge bonds per atom, a homonuclear single covalent bond energy of 45.0 kpm would seem appropriate, assuming the discrepancies noted for carbon and perhaps silicon to be absent here. However, the value calculated from the $E = CrS$ equation is 49.1, which is appreciably higher. Back-calculation from the experimental atomization energies of the gaseous halides gives the following values: $GeCl_4$, 48.5; $GeBr_4$, 52.0; GeI_4, 49.2; averaging 49.9 (all based on the X''' energies of the halogens). Therefore the value 49.1 is tentatively adopted.

ARSENIC

The bond length in As_4 is 2.43 corresponding to a nonpolar covalent radius of 1.22 Å. However, a "tetrahedral" radius of 1.18 is reported, and interpolation of the Z/r^3 function between germanium and bromine gives 1.19 Å, which is the value adopted here.

The electronegativity of arsenic is 3.90. The experimental heats of formation are 72.3 for monatomic As(g) and 34.4 for As_4(g) from which the atomization energy of As_4 is 254.8 kpm. Atomization of As_4 requires breaking six bonds per molecule, giving an As—As bond energy of 42.5 kpm. However, the bond length is 2.43 Å instead of the covalent radius sum of 2.38 Å. Correction to the latter gives 43.4 for the As' homonuclear single covalent bond energy.

The As''' bond energy calculated from the $E = CrS$ equation is 52.0 kpm. Halfway between 52.0 and 43.4 is 47.7 kpm, the value of the As'' energy.

SELENIUM

The bond length in elemental selenium is 2.32 Å corresponding to a nonpolar covalent radius of 1.16 Å, which is also the value interpolated between germanium and bromine using the Z/r^3 function. The electronegativity is 4.21. Although there is no known application at present, the calculated homonuclear single covalent bond energy for Se''' from the $E = CrS$ equation is 54.6 kpm. The Se' energy has been reported as about 44 kpm. This would correspond to a Se'' energy of 49.3 kpm.

It is interesting that the standard heat of formation of Se(g) is 49.2 kpm. If atomization of selenium from its standard state is equivalent to the breaking of one covalent bond per atom, this suggests that in selenium as in sulfur (S_8) the Se'' bond energy applies. Further, the heat of formation of

$Se_2(g)$ is 34.9, from which the energy of the Se—Se bond in this molecule is 63.5 kpm. The bond length is 2.15 Å, compared to the covalent radius sum of 2.32 Å. If 49.2 is the correct energy for Se″, then the bond in Se_2 cannot possibly be a double bond. The calculated multiplicity factor is 1.20 instead of the olefinic factor of 1.50. Data are lacking for testing the validity of these energies.

BROMINE

The bond length in Br_2 is 2.28 Å from which the nonpolar covalent radius of 1.14 Å is obtained. The relative compactness electronegativity is 4.53. The experimental dissociation energy of Br_2 is 46.1 kpm which is adopted for the Br′ homonuclear single covalent bond energy.

Application of the $E = CrS$ equation gives a value of 58.0 for the Br‴ parameter. The Br″ energy is then 52.0 kpm. The 58.0 value is supported by back-calculation from the atomization energies of a number of gaseous bromides, giving results as follows: $BeBr_2$, 54.6; BBr_3, 55.0; $AlBr_3$, 61.1; $GeBr_4$, 64.0; and $SnBr_4$, 53.7. The average is 57.5 kpm.

RUBIDIUM

The diatomic molecules of rubidium that exist in small concentration in rubidium vapor have a bond length of 4.32 Å corresponding to a nonpolar covalent radius of 2.16 Å. The electronegativity is 0.36. The experimental dissociation energy of Rb_2 at 25° is 12.4 kpm. The homonuclear single covalent bond energy calculated by the equation $E = 16.1\, rS$ is 12.4 kpm.

STRONTIUM

The nonpolar covalent radius of strontium is taken as 1.91 Å and the electronegativity as 1.06. By the $E = CrS$ equation, the homonuclear single covalent bond energy is calculated as 32.6 kpm. No direct experimental verification is possible, but back-calculations can be made from the atomization energies of the solid halides. Results are: SrF_2, 32.1; $SrBr_2$, 32.6; SrI_2, 33.4; average, 32.7 kpm. Therefore the value 32.6 is tentatively adopted.

CADMIUM

The nonpolar covalent radius of cadmium extrapolated from iodine and tin using the Z/r^3 function is 1.46 Å. The electronegativity is taken as 2.59. From the $E = CrS$ equation, where $C = 8.6$ corresponding to $n = 5$, the homonuclear single covalent bond energy of cadmium is calculated to be 32.5 kpm. These values give only rough agreement between calculated and experimental atomization energies for the limited number of cadmium data available. However, agreement would be much improved if the atomization energy of cadmium were 42.7 instead of 27. Problems obviously exist here which have not yet been solved.

INDIUM

The nonpolar covalent radius of indium also extrapolated from iodine and tin is 1.43 Å. The electronegativity is 2.84. From the $E = CrS$ equation, the homonuclear single covalent bond energy of indium is calculated to be 35.0 kpm. An attempt to verify this value by calculation of the atomization energies of the gaseous halides gives results that are about 13 kpm too low for chloride, bromide, and iodide. This suggests that if the atomization energy of indium were about 45 kpm instead of the reported 58, excellent agreement would be observed between the experimental and calculated atomization energies.

A value of only 25.3 kpm is calculated from the experimental heats of formation of 58.2 for In(g) and 91.0 for $In_2(g)$.

TIN

The experimental bond length in gray tin is 1.40 Å. The electronegativity is 3.09. The homonuclear single covalent bond energy calculated from these data and the $E = CrS$ equation is 37.2. The experimental atomization energy of gray tin is 72.2 kpm corresponding to a bond energy of 36.1 kpm. Too few data are available to justify a selection. Either value gives satisfactory agreement between experimental and calculated bond energies in tin compounds.

In the +2 state the presence of the inert pair is believed to reduce the electronegativity of tin to a value of about 2.31. The above data therefore probably apply only to tetravalent tin.

ANTIMONY

The nonpolar covalent radius of antimony interpolated from the Z/r^3 function of tin and iodine is 1.38 Å. The electronegativity is 3.34. From the $E = CrS$ equation, the homonuclear single covalent bond energy is calculated to be 39.6 kpm. This would be the Sb‴ value. An Sb′ value of 32.0 has been reported. The Sb″ value would then be 35.8 kpm.

TELLURIUM

In literature compilations, the nonpolar covalent radius of tellurium has been reported as 1.37 for the "normal" radius and 1.32 for the "tetrahedral" radius. By interpolation between tin and iodine the value 1.35 Å is obtained. The electronegativity of tellurium is 3.59. By use of the $E = CrS$ equation the homonuclear single covalent bond energy for Te‴ is calculated as

41.7 kpm. The Te′ energy has been reported as about 34 kpm which would make the Te″ energy about 38 kpm.

IODINE

The bond length in I_2 molecule is 2.66 Å from which the nonpolar covalent radius is 1.33 Å. The relative compactness electronegativity is 3.84. The dissociation energy of I_2 is 36.1 kpm, which is the value for the homonuclear single covalent bond energy for I′. The value for I‴ is calculated from the $E = CrS$ equation to be equal to 43.9 kpm which would correspond to an I″ value of 40.0 kpm. Fewer examples of its application exist, but the I‴ value is supported by back-calculations from gaseous BeI_2 (43.2) and AlI_3 (42.1 kpm).

CESIUM

The bond length in the diatomic molecule of $Cs_2(g)$ is 4.70 Å corresponding to a nonpolar covalent radius of 2.35 Å. The electronegativity is 0.28. The dissociation energy of $Cs_2(g)$ is experimentally determined as 10.7 kpm. The $E = CrS$ equation using the above data gives 10.6 kpm.

BARIUM

The nonpolar covalent radius of barium is taken as 1.98 Å and the electronegativity 0.78. Application of the $E = CrS$ equation gives 24.9 kpm for the homonuclear single covalent bond energy. An empirical relationship between radii and bonding energies in diatomic molecules and body-centered cubic metallic lattice described in the following chapter gives for barium the homonuclear single bond energy of 25.7 kpm. These values give good agreement between experimental and calculated atomization energies of barium compounds.

Other Major Group Elements

For a number of other elements which are outside the present scope of the $E = CrS$ relationship, it is possible to make reasonable estimates of their homonuclear single covalent bond energy by back-calculation from experimental atomization energies. This procedure without outside support has, of course, a circular character that may reasonably impair confidence, since it involves using experimental bond energies to calculate energies which are used only for calculating bond energies. Nevertheless, it is significant that essentially the same values are calculated from a series of different compounds. The following data are tentative and their basis is fully described to permit individual judgment as to their possible validity.

MERCURY

Uncertainty as to structural parameters of solid mercury compounds requires reliance on data for the gaseous dihalides to estimate the homonuclear single covalent bond energy of mercury. These data are quite inadequate. In $HgF_2(g)$, the covalent energy contribution is too small for use in back-calculation of bond energy. The electronegativity of mercury is 2.93 and the nonpolar covalent radius 1.49 Å. If the halogen is assumed to use its single prime energy in its bonds to mercury, then the following Hg—Hg energies are back-calculated: $HgCl_2$, 7.5; $HgBr_2$, 7.1; HgI_2, 11.3; average, 8.6 kpm. Use of this average to calculate atomization energies gives the following results (calculated—experimental): HgF_2, 145.0—122.7; $HgCl_2$, 110.4—107.9; $HgBr_2$, 91.8—88.5; HgI_2, 64.8—69.6 kpm.

THALLIUM

Data are available only for the (I) state of thallium—the halides. The electronegativity of thallium (I) is 1.89 and its covalent radius is 1.48 Å. From the four solid halides can be back-calculated an average Tl—Tl energy of 17.8 kpm. Use of this value in calculating the atomization energies of the halide salts gives the following results (calculated—experimental): TlF, 141.5—140.1; TlCl, 122.0—121.7; TlBr, 111.4—111.5; TlI, 97.1—98.8 kpm.

LEAD

The electronegativity of lead (II) is 2.38 and that of lead (IV) 3.08. The covalent radius is 1.47 Å. Numerous data for lead halides are available. Back-calculation of the Pb—Pb energy from the solid halides gives the following results: PbF_2, 31.2; $PbCl_2$, 17.4; $PbBr_2$, 20.3; PbI_2, 24.1 kpm. The gaseous mono- and dihalides appear to involve the triple prime energies of the halogen atoms, which by back-calculation give the following Pb—Pb energies: Pb—Cl, 23.4; $PbCl_2$, 19.7; PbBr, 19.7; $PbBr_2$, 19.9; PbI, 19.0; PbI_2, 20.7 kpm. The value for $PbF_2(c)$ was omitted in averaging to give the value 20.5. This is the adopted value.

The atomization energy of $PbF_4(g)$ is clearly too low to involve the F''' energy. It appears that the X' energy is applicable to the tetrahalides. Back-calculation of the Pb—Pb energy from these gives the following results: $PbCl_4$, 20.6; $PbBr_4$, 18.8; and PbI_4, 21.4; average 20.3, practically the same as for the Pb(II) compounds. It is therefore assumed correct and that the Pb—Pb energy does not depend on the oxidation state or number of bonds formed by the lead. This is of course consistent with earlier results.

BISMUTH

Bismuth has an electronegativity value of 3.16 and a covalent radius of

1.46 Å. Data on arsenic and antimony halides, although generally poor, suggest that the triple prime energy of the halogen should be applicable to the bismuth trihalides. Back-calculating from experimental atomization energies leads to an average value of 30.4 kpm for the homonuclear single covalent bond energy of bismuth. Application of this value to the three trihalides for which data are available gives the following results: (calculated— experimental): $BiCl_3$, 201.3—200.0; $BiBr_3$, 167.7—168.1; and BiI_3, 129.6— 129.9 kpm.

Transitional Elements

Bond energy calculations for the transitional elements are complicated by uncertainties as to crystal field effects, electronegativities, covalent radii, and other factors. However, $3d$ electron effects can be considered minimized in compounds of titanium (IV), manganese (II), and copper (I) and silver (I).

TITANIUM

Data for the gaseous tetrahalides are available. The F'' energy appears appropriate for TiF_4 but the X' energy for the other halides. From these a Ti—Ti energy is back-calculated with the following results: $TiCl_4$, 13.0; $TiBr_4$, 11.1; TiI_4, 17.3; average 13.8 which is adopted. When this value is applied to calculating the atomization energy of titanium (IV) compounds, the following data are obtained for the halides (calculated—experimental atomization energies): TiF_4'', 558.4—558.8; $TiCl_4$, 412.0—411.3; $TiBr_4$, 355.2—350.0; TiI_4, 277.2—283.0. The value can also be applied to solid TiO_2, giving a calculated atomization energy of 459.6 to be compared with the experimental value of 458.0 kpm.

The covalent radius is 1.32 Å and the electronegativity 1.40.

MANGANESE

The covalent radius is 1.17 Å and the electronegativity 2.07. Back-calculation of the Mn—Mn energy from atomization energies of the solid compounds gives the following results: MnF_2, 31.9; $MnCl_2$, 30.6; $MnBr_2$, 30.7; MnI_2, 35.7; MnO, 39.6; MnS, 31.6; average value of 33.4 kpm is adopted.

COPPER

The nonpolar covalent radius of copper extrapolated by the Z/r^3 function from germanium and bromine is 1.35 Å. The electronegativity is 2.60. Back-calculation from atomization energies experimentally determined for the solid (I) halides gives the following values of the Cu—Cu energy: CuF, 15.6; CuCl, 15.5; CuBr, 16.9; CuI, 20.2; average value of 17.1 is adopted.

TABLE 4-1

ATOMIC PROPERTIES FOR BOND ENERGY CALCULATIONS

Element	r_c (Å)	S^a	Log S	ΔS_i^b	$\Delta Hf°$(g) (kpm)	Homonoculear bond energy (kpm)		
						E'''	E''	E'
H	0.32	3.55	0.5502	3.92	52.1	104.2	—	—
Li	1.34	0.74	− 0.1308	1.77	38.4	28.3	—	—
Be	0.91	2.39	0.3784	3.22	78.3	62.2	—	—
B	0.82	2.93	0.4669	3.56	134.5	68.4	—	—
C	0.77	3.79	0.5786	4.05	171.3	83.2	—	—
N	0.74	4.49	0.6522	4.41	113.0	94.8	67.0	39.2
O	0.72	5.21	0.7168	4.75	59.6	99.4	66.7	34.0
F	0.71	5.75	0.7597	4.99	18.9	100.3	69.1	37.8
Na	1.54	0.70	− 0.1549	1.74	25.8	18.0	—	—
Mg	1.38	1.56	0.1931	2.60	(47)	34.6	—	—
Al	1.26	2.22	0.3464	3.10	78.0	45.0	—	—
Si	1.17	2.84	0.4533	3.51	108.9	53.4	—	—
P	1.10	3.43	0.5353	3.85	79.8	60.7	55.9	51.1
S	1.04	4.12	0.6149	4.22	66.6	69.0	62.1	55.0
Cl	0.99	4.93	0.6928	4.62	29.1	78.6	68.4	58.2
K	1.96	0.42	− 0.3768	1.35	21.3	13.2	—	—
Ca	1.74	1.22	0.0864	2.30	42.5	34.2	—	—
Zn	1.30	2.98	0.4742	3.58	31.	43.4	—	—
Ga	1.26	3.28	0.5159	3.77	66.2	46.3	—	—
Ge	1.22	3.59	0.5551	3.94	90.0	49.1	—	—
As	1.19	3.90	0.5911	4.11	72.3	52.0	47.7	43.4
Se	1.16	4.21	0.6243	4.27	49.2	54.6	49.3	44.0
Br	1.14	4.53	0.6561	4.43	26.7	58.0	52.0	46.1
Rb	2.16	0.36	− 0.4437	1.25	19.6	12.4	—	—
Sr	1.91	1.06	0.0253	2.14	39.	32.6	—	—
Cd	1.46	2.59	0.4133	3.35	27.	32.5	—	—
In	1.43	2.84	0.4533	3.51	(45.2)	35.0	—	—
Sn	1.40	3.09	0.4900	3.66	72.2	37.2	—	—
Sb	1.38	3.34	0.5237	3.80	62.7	39.6	35.8	32.0
Te	1.35	3.59	0.5551	3.94	45.5	41.7	38.	34.
I	1.33	3.84	0.5843	4.08	25.5	43.9	40.0	36.1
Cs	2.35	0.28	− 0.5528	1.10	18.7	10.7	—	—
Ba	1.98	0.78	− 0.1079	1.93	42.	24.9	—	—
Hg(II)	1.49	2.93	0.4669	3.59	14.7	8.6	—	—
Tl(I)	1.48	1.89	0.2765	2.85	43.6	17.8	—	—
Pb(II)	1.47	2.38	0.3766	3.21	46.8	20.5	—	—
Pn(IV)	1.47	3.08	0.4886	3.69	46.8	20.5	—	—
Bi(III)	1.46	3.16	0.4997	3.74	49.5	30.4	—	—
Ti(IV)	1.32	1.40	0.1461	2.48	113.0	13.8	—	—
Mn(II)	1.17	2.07	0.3160	2.99	67.	33.4	—	—
Cu(I)	1.35	2.60	0.4150	3.36	81.	17.1	—	—
Ag(I)	1.50	2.57	0.4099	3.33	68.	15.7	—	—

^{a} S is the electronegativity.
^{b} ΔS_i is the change in electronegativity that would accompany acquisition of unit charge.

SILVER

The nonpolar covalent radius of silver extrapolated by the Z/r^3 function from tin and iodine is 1.50 Å. The electronegativity is 2.57. Back-calculation from the experimental atomization energies of the solid halides gives the following values of the Ag—Ag energy: AgF, 17.6; AgCl, 14.1; AgBr, 15.8; AgI, 15.2; the average of 15.7 adopted.

Summary

The numerical data for this chapter are summarized in Table 4-1, together with other necessary or helpful data for bond energy calculations.

From the above details it should be clear that experimental data have been used wherever possible to validate or provide the selection of fundamental atomic properties to be used for the calculation of bond energies. Although in some cases the choice of atomic property was somewhat arbitrary, in general they are well founded and should be reasonably reliable. The interrelationship between electronegativity and homonuclear single covalent bond energy is sufficiently quantitative to justify the faith that both quantities are based on the same nuclear–electronic interactions, and for most of the elements are experimentally verifiable through quantitative measurement of the homonuclear bond energy.

It is only fair to point out that even if the electronegativity and bond energy were pulled out of a hat, their quantitative applicability to the heretofore very difficult problem of calculating bond energies in heteronuclear bonds would be extremely impressive. The data of the following chapters should supply reinforcement to all the concepts on which the calculations are based, as well as to the fundamental properties of Table 4-1

Chapter 5

ATOMIZATION AND STATES OF AGGREGATION OF THE ELEMENTS

Metals

Quantitative data for explaining the evident preference of certain elements for the metallic state are usually unavailable. The alkali metals offer an exception, making it relatively easy to understand why sodium, for example, exists as a solid rather than as Na_2 gas molecules at ordinary temperatures. The latter molecules do occur in small concentration in the vapor. Their dissociation energy is 18.0 kpm, whereas the atomization energy of sodium is 25.9 kpm from the metallic crystal. But since dissociation of a mole of diatomic molecules produces 2 moles of sodium atoms, the atomization energy of Na_2 per mole of atoms is only 9.0 kpm. Thus the energy gain, or increase in stability of the metallic state over the diatomic gas, is 16.9 kpm of sodium atoms. Quailtatively, this difference may be explained as the result of the advantage of spreading the valence electrons among all the available orbital vacancies instead of forcing them to concentrate in pairs, one pair in each particular internuclear region.

Quantitatively no theoretical relationship between the nonpolar covalent bond energy and the metallic atomization energy has been discovered. The latter, of course, is the measure of the total bonding energy in the metallic state. Empirically, however, the dissociation energy of the diatomic gas molecules of the alkali metals, when corrected to the metallic bond length which is always substantially larger, has been found simply related to the atomization energy of the body centered cubic lattice:

$$E_c = \frac{E_{at}R_M}{1.83R_c} \tag{5-1}$$

Table 5-1 permits a comparison of the experimental dissociation energies of

59

the alkali metals with the values calculated from metallic state data by this equation.

TABLE 5-1

HOMONUCLEAR SINGLE COVALENT
BOND ENERGIES CALCULATED FROM METAL PROPERTIES

Metal	$E_{atomization}$	R_M	R_c	Covalent bond energy (kpm)	
				Calc.	Exp.
Lithium	38.4	3.10	2.68	24.1	26.5
Sodium	25.9	3.80	3.08	17.4	18.0
Potassium	21.4	4.70	3.92	13.9	13.2
Rubidium	19.6	4.96	4.32	12.2	12.4
Cesium	18.7	5.34	4.70	11.5	10.7
Barium	42	4.44	3.96	25.6	27.5

Nonmetals

Bond energy calculations afford ready explanations of the states of aggregation of nonmetallic elements, clearing up numerous "anomalies" that have been puzzling to chemists for many years. In general, the spectacularly abrupt change in crossing period 2 from lithium to neon is closely associated with the common cause of these anomalies. The melting point, for example, rises steadily from lithium (180.5°), beryllium (1283°), boron (2000°), to carbon (about 3500°), and then falls to below −200° for nitrogen, oxygen, and fluorine. The usual practice is to recognize that the latter three elements exist as diatomic molecules having only very weak van der Waals interactions, which account for their physical properties of very low melting point and high volatility. Fluorine, of course, presents no problem. Its atoms are limited by their electronic structure to one single covalent bond each. Therefore diatomic molecules F_2 represent the highest degree of condensation possible for them, except through the very weak van der Waals forces that exist among such small molecules. But the question of *why* is not answered for nitrogen and oxygen by the customary statement that these elements readily form multiple bonds. For the question really is: Why do nitrogen and oxygen form multiple bonds in preference to single bonds?

M5 ELEMENTS

It has long been known that double bonds formed by carbon have an energy of about 144 kpm, not twice as great as the single bond energy of 83 kpm. Triple bonds formed by carbon have an energy of about 190 kpm, certainly not three times as great as the single bond energy nor even 1.5 times as great as the double bond energy. These differences are readily interpreted qualitatively in terms of the increasing difficulty of concentrating additional electrons within the internuclear region because of their mutual repulsion. Then why does nitrogen show an opposite tendency?

As described in detail in Chapter 3, the advent of the lone pair appears responsible for the abrupt change in character from carbon to nitrogen. We may here examine quantitatively the alternatives to triple-bonded nitrogen, N_2. For nitrogen might associate through single covalent bonds, each atom being joined to three others. Another possibility would be long chains alternating single and double bonds: $-N=N-N=N-N=$. A combination of these two would also be possible. Or, why not benzene-like N_6 rings?

Let us consider first the singly bonded polymer. The single bond energy normally is the N' energy of 39.2 kpm. Atomization of the polymer would require the equivalent of breaking 1.5 bonds per nitrogen atom, since each break releases two atoms from the bond. The atomization energy of the polymer would therefore be only $39.2 \times 1.5 = 58.8$ kpm, far lower than the 113 kpm of nitrogen atoms from N_2 molecules. The energy of polymerization would then be about $+54$ kpm. Removal of gaseous molecules of N_2 to form a solid polymer would certainly involve a decrease in entropy since the polymeric form would have much less disorder than the gas. Therefore the free energy change for polymerization of N_2 molecules to a solid would have a larger positive value than $+54$. No further question should arise as to why nitrogen does not exist as a single-bonded polymer.

Next, we may consider the long chains of alternate single and double bonds. A double bond energy for nitrogen would involve the factor 1.50 and the single bond parameter for N'', 67.0, as well as a shorter than single bond length. The double bond energy is calculated to be 109.0 kpm. Conjugation in the polymer chains described would somewhat lengthen and weaken the double bonds and shorten and strengthen the single bonds, but as a reasonable approximation we may ignore this effect and simply take the average of the double and single bond energies to give the atomization energy of the chain polymer: 74.1 kpm. This is still much lower than the 113 for N_2, making the energy of polymerization of N_2 to the chain polymer in the neighborhood of $+35$ kpm. The free energy change would be larger than this, since again the entropy would decrease. The chain model thus shows little promise of successful competition with the N_2 molecule.

The hypothetical benzene-like ring compound, N_6, is not quite so easily treated, but at least we can place an ample upper limit on the bond energy if we state that it certainly cannot exceed that of an $N=N$ double bond.

The carbon–carbon bonds in the benzene ring are longer and weaker than $C{=}C$ double bonds, about 122 kpm compared to 144 kpm, so the above reasoning seems well based. Therefore the atomization energy of the N_6 rings, which would be equivalent to the energy of one of the NN bonds since six atoms are liberated by the breaking of six bonds, must be substantially lower than that of N_2 molecules—somewhere between 74 and 109 kpm, and probably nearer the former. Again it becomes clear why the preferred form of nitrogen is the N_2 molecule. In all these possibilities, the decisive factor is the greater *single* bond energy associated with multiplicity of nitrogen.

Having rationalized the gaseous state of nitrogen, we must now explain why the remaining elements of this group, beginning with phosphorus, are not also gaseous diatomic molecules. It is not fair to imply that the other elements, especially phosphorus, are incapable of multiple bonding. When phosphorus is vaporized, the vapor species is P_4, in which four phosphorus atoms are held together by single covalent bonds. However, when this vapor is heated above 800°, it breaks down to P_2 molecules which are very stable. The bond length is 1.89 Å compared to 2.20 Å for the nonpolar covalent radius sum, indicating multiplicity since no polarity can be present. The dissociation energy is found to be 125.1 kpm, experimentally. If the bond is assumed to be an ordinary triple bond, then use of the triple bond factor permits calculation of the P''' single bond energy parameter of 60.7, which as shown in the preceding chapter is precisely the value obtained from the $E = CrS$ equation. In other words, it is completely consistent with the hypothesis that the bond in P_2 is an ordinary triple bond like that in N_2.

It is important to notice that here the difference between E' and E''' parameters is much smaller than for nitrogen. This fact is very significant. It is the basis of an explanation of the solid state of phosphorus. Neglecting very minor bond length changes, we may calculate the atomization energy of P_4, the usual form of phosphorus in its vapor at lower temperatures and in the white modification, as equivalent to 1.5 times the single bond energy. This is because three bonds must be broken to liberate one atom, but each broken bond liberates two atoms from that bond. Then $1.5 \times 51.1 = 76.7$ kpm of atoms. Atomization of the P_2 molecule, on the other hand, would require only half the dissociation energy of 125.1 kpm per mole of atoms or 62.6 kpm. The enthalpy of polymerization of P_2 to P_4 is therefore $62.6 - 76.7 = -14.1$ kpm of atoms. This accounts for the ready condensation of the smaller to the larger molecules.

White phosphorus is still not the ideal arrangement of phosphorus atoms because of the strain of 60° bond angles in P_4. This strain is eliminated in the more highly condensed forms of phosphorus.

Adequate data for a similar study of the heavier elements of this group do not exist, but the tentative conclusions based on reasonable estimates of the different bond parameters of these elements are similar. That is, in each of these elements, the difference between E' and E''' parameters is no greater than in phosphorus, which means that the multiply bonded states would not be favored energetically.

M6 ELEMENTS

The preference of oxygen for O_2 molecules rather than a higher degree of association is of course similar to that of nitrogen. The divalence of oxygen, however, limits the possibilities more than for nitrogen. One can picture only long chains or rings of oxygen atoms, singly bonded to each neighbor. The single bond energy which is the O' parameter is only 34.0 kpm. This is the atomization energy of the polymer, since each broken bond liberates two atoms from the bond. On the other hand, the atomization energy of the O_2 molecule is half its dissociation energy of 119.2, or 59.6 kpm of atoms. The enthalpy of polymerization would then be $59.6 - 34.0 = +25.6$ kpm of atoms. Polymerization of a gas decreases the disorder and thus reduces the entropy. Therefore the free energy of polymerization would be larger than $+25$ kpm of atoms. This accounts for the existence of free oxygen as O_2 rather than chain or ring polymers. Again it is the partial reduction in the "lone pair weakening" associated with the multiple bond that causes the double bond to be favored. If it were not for this effect, then polymeric oxygen would be favored because then a double bond would not be as strong as two single bonds.

Oxygen does exist, however, as ozone, O_3, which is very unstable with respect to O_2. A molecule of ozone is bent at an angle of about 120°. This suggests that only one of the lone pairs of the central oxygen is not in use, giving three locations for electrons about this oxygen. We can then picture the molecule as an average of O—O=O and O=O—O, where the single bond must be a coordination bond. The question of which single bond energy parameter is applicable is answered by averaging the O' and the O'' parameters to give 50.4. Then the single (coordination) bond energy at the observed bond length is $(50.4 \times 1.44)/1.28 = 56.7$ kpm. The double bond energy is $(50.4 \times 1.50 \times 1.44)/1.28 = 85.1$. The sum, 141.8, agrees reasonably well with the experimental value of 143.5 kpm. It corresponds to an enthalpy of formation of ozone of $+35.8$ kpm, to be compared to the experimental value of $+34.1$. This calculation shows again the advantage of diatomic molecules of O_2 that results from the reduction of "lone pair weakening" in the double bond.

So much for why oxygen is a gas. What about sulfur? Why should it not also exist as gaseous molecules, S_2? Once more one may observe, as in the comparison of nitrogen with phosphorus, that the difference between E' and

E''' parameters is much less for the heavier elements than for the first member of the group. Again it is unfair to imply that sulfur cannot form double bonds like those in O_2. For when sulfur vapor, initially S_8, is heated its molecules grow smaller until the predominant species between 500 and 1900° is S_2. There appears to be little reason to doubt that these are analogous to O_2 molecules. The bond is appreciably shorter than in S_8, and the molecules are paramagnetic indicating two unpaired electrons per molecule as in O_2. The experimental bond energy is 102.5 kpm and the bond length 1.89 Å. As indicated in the preceding chapter, one can calculate from these data an S'' energy parameter of 62.1 kpm. Since the double bond energy of 102.5 is less than twice the S' energy of 55.0, enthalpy alone would favor polymerization over dimerization in sulfur. However, this might be offset by the unfavorable entropy change associated with polymerization, since the enthalpy difference is not great. But the enthalpy difference is actually appreciably greater than would be judged from the S' energy of 55.0. The favored polymeric form is S_8 in which the bond energy is greater by 8.5 kpm of atoms.

The atomization energy of sulfur is 66.6 kpm of atoms. The vaporization energy of S_8 from the crystal to S_8 gas molecules at 25° is 24.5 kpm. The average bond energy in S_8 thus comes out to be 63.5 kpm, corresponding nearly to the S'' energy instead of the expected S' energy. Why this is so is not yet known. But it is probably tied in with the special advantages of S_8 rings over some other type of association. Possibly the S_8 geometry somehow appreciably reduces the lone pair bond weakening expected for S—S single bonds, and more so than permitted by other geometries. Some such explanation must be reasonable, for how else can one account for the spontaneous change of S_6 rings, and all linear fragments of sulfur chains, to S_8 rings at 25°? Whatever the explanation, it provides a bonus in stability to the polymeric single-bonded sulfur over the gaseous double-bonded dimer, making the total energy of polymerization from the dimer − 15.3 kcal per gram-atom. This is why sulfur is a solid instead of being gaseous like oxygen. For if the same single bond parameter is applicable, then two single bonds must always be stronger than one double bond.

Data are lacking for selenium and tellurium, but similar conclusions are probably reasonable. There may also be some grounds for suspicion, as indicated in the preceding chapter, that stable double bonds may be less possible for the larger atoms.

Although a very self-consistent picture has been presented in the preceding discussion, the explanations cannot be regarded as completely satisfying until a quantitative accounting for the "lone pair" effects can be accomplished. It is quite evident that these effects are much larger in the smaller atoms of each group. This difference constitutes the source of the anomalies of the periodic table discussed above.

Chapter 6

APPLICATIONS TO INORGANIC MOLECULES

I. OXIDES, OXYACIDS, AND THIOANALOGS

This chapter is deliberately restricted to a consideration of inorganic *molecules* (including giant polymers). This means that all the many non-molecular solids formed by combination of oxygen with metallic elements are omitted from the present discussion. A large variety of molecular oxides exists, however, several of which offer useful examples of interesting applications of bond energy considerations. An attempt will be made to focus on some of the problems these oxides have presented individually, and to explain some of the facts of chemistry by which these problems can be made to vanish.

WATER

The magnitude of the conversion factor, 332, used in calculating heteronuclear bond energies is so great that, in general, the ionic energy is disproportionately large. The homonuclear nonpolar covalent bond energy parameter enters only into the calculation of the covalent energy contribution. The effect of differences among E', E'', and E''' parameters therefore becomes less and less significant the more polar the bond.

The water molecule affords an excellent example of the great importance of bond polarity. Let us suppose that its bonds were completely nonpolar. If so, their bond length would equal the covalent radius sum and their bond energy would be $(104.2 \times 34.0)^{\frac{1}{2}} = 59.5$ kpm. This corresponds to an atomization energy of 119.0 kpm of water. To atomize 2 H + O from their standard states of gaseous H_2 and O_2 would require $104.2 + 59.6 = 163.9$ kcal. The difference or standard heat of formation of $H_2O(g)$ would be $163.9 - 119.0 = +44.9$ kpm. The free energy change would be even more positive, by about 3 kpm, since the formation of $H_2O(g)$ produces a decrease in the number of gas molecules and therefore in entropy.

In summary, if OH bonds were nonpolar water would be thermodynamically unstable. Not only would hydrogen fail to burn, but heat could be extracted from water by decomposing it to its elements! This not only shows the importance of bond polarity, but also rather dramatically emphasizes the advantage of bond multiplicity for oxygen. For two nonpolar bonds to oxygen are much weaker than one double bond in O_2.

In a real water molecule, on the other hand, the electronegativity difference between hydrogen and oxygen results in partial charges of 0.12 on hydrogen and -0.25 on oxygen. Their average leads to 0.18 for t_i and, consequently, 0.82 for t_c. The hydrogen bond energy parameter is corrected, as will be explained in Chapter 8, by the factor $(1.00 - \delta_H) = 0.88$, to the value 91.7. The single bond O' parameter of 34.0 is applicable. The geometric mean of these is 55.8. Polarity shrinks the bond from the covalent radius sum of 1.04 to 0.96 Å. The covalent energy contribution is therefore

$$t_c E_c = \frac{0.82 \times 55.8 \times 1.04}{0.96} = 49.6 \text{ kpm}$$

The ionic energy contribution is

$$t_i E_i = \frac{0.18 \times 332}{0.96} = 62.2 \text{ kpm}$$

The total energy of the bond is then the sum 111.8. Twice this is the atomization energy of water, 223.6 kpm. This agrees well with the experimental value of 221.6 kpm. The calculated heat of formation is -59.8 kpm, to be compared with the experimental value of -57.8 kpm. An important point to recognize in these details is that although the ionic coefficient is only 0.18, the ionic contribution is 56% of the total bonding energy. The advantage of a double bond to oxygen is thus outweighed by the effect of polarity and water becomes a very stable molecule. See Figs. 6-1 and 6-2.

POLYWATER

For a number of years, evidence has been accumulating that a strange form of water results when water vapor condenses in very tiny fused quartz capillaries. This form, called "polywater,"[1] has recently been carefully studied by infrared and Raman spectroscopy. Although thus far only available in minute quantities it seems to be very stable and differs from ordinary water in having a density of approximately 1.4 and in giving no evidence of the presence of ordinary O—H bonds. In fact the spectrum is not that of any other known substance. The authors of this study believe the material to be a highly polymeric substance in which each oxygen atom is bridged to three other oxygen atoms through hydrogen. These "hydrogen bonds" (see p. 116) resemble the bridge in the bifluoride ion in being very short and

[1] E. R. Lippincott, R. R. Stromberg, W. H. Grant, and G. L. Cessac, *Science* **164**, 1482 (1969).

strong and in having the hydrogen atom centered between the two oxygen atoms. Each O—H bridge is considered to be of 2/3 bond order and of energy 30–50 kpm. They estimate a bond length of 1.15 Å. They suggest the possibility of the substance being a polyelectrolyte, where a certain number of protons form hydronium ions, leaving a number of negative charges on the anion polymer.

Since the formation of such bridging is not the normal behavior of hydrogen bearing partial positive charge (see Chapter 8), it is interesting to speculate that perhaps enough protons must leave to reduce the charge

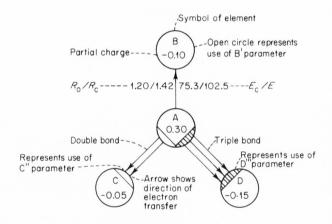

Fig. 6-1. Schematic representation of bonding in an imaginary molecule.

on hydrogen to a zero or negative value. Calculations show that this would require at least the loss of one proton for every three water molecules, which would bring the charge on hydrogen to approximately zero, and suggest a possible formula such as $nH_3O^+(H_5O_3^-)n$ for the polywater. However, as Lippincott suggests, $nH_3O^+(H_3O_2^-)n$ might be better, and in fact this would correspond to partial charges in the anion of -0.06 on hydrogen and -0.40 on oxygen. These resemble the charges calculated for the bifluoride ion: hydrogen -0.04 and fluorine -0.48. Thus there would be not only strong bonding within the polymeric anion but also strong protonic bridging between the hydronium ions with 0.35 charge on hydrogen and the anion oxygens of -0.40 charge. Polywater would thus consist of a highly branched chain structure of polyanion with a hydronium ion attached to every other anion oxygen by a protonic bridge.

As will be discussed in Chapter 8, hydrogen bearing partial negative charge commonly appears to form bonds that in effect are nonpolar so far as bond energy calculations are concerned. It is therefore not necessary to know exactly what charge is on hydrogen as long as it is negative in order to calculate the energy of the O—H—O bridge bonds. The geometric mean of 104.2 for hydrogen and 34.0 for oxygen is 59.5. The covalent radius sum is 1.04 Å and the bond length of 1.15 Å and the bond order of 0.67 have been proposed. (The bond order recognizes the use of only two pairs of electrons in formation of three bonds from each oxygen to hydrogen.) On this basis, the energy is

$$E = \frac{0.67 \times 59.5 \times 1.04}{1.15} = 35.9 \text{ kpm}$$

This value may be recognized as within the estimated range of 30 to 50 kpm.

Possibly this kind of hydrogen bridging is similar to that observed in the alkali metal hydroxide hydrates where, for example, KOH . H_2O appears to be $K^+(HOHOH^-)$. In this anion, the partial charges are calculated to be -0.07 on hydrogen and -0.40 on oxygen.

Further experiments may well disclose facts contrary to the above speculations.[2] The subject is discussed here to illustrate how the methods of this book may be applied in the analysis of a new situation.

HYDROGEN PEROXIDE

The well-known instability of hydrogen peroxide with respect to water and oxygen is easily explained on the basis of bond energies. The number of OH bonds in a molecule of hydrogen peroxide is the same as in a molecule of water and their energies are almost identical. However, in hydrogen peroxide there is the extra feature of two oxygen atoms united by a single covalent bond which utilizes the O' parameter of 34.0. A double bond between oxygen atoms utilizes the O'' parameter of 66.7 and also the multiplicity factor 1.50. It is therefore more than twice as strong. The calculated heat of formation of $H_2O_2(g)$ is practically identical with the experimental value of -32.6 kpm (see Fig. 6-2). For the reaction,

$$2 H_2O_2(g) \rightarrow 2 H_2O(g) + O_2(g)$$

the enthalpy change must be the difference between the standard heats of formation of the products and the reactants: $(2 \times -57.8) - (2 \times -32.6) = -50.4$ kcal, or -25.2 kpm of hydrogen peroxide.

It is important to recognize that the concept of thermodynamic stability with respect to the chemical elements in their standard states, although

---

[2] D. L. Rousseau and S. P. S. Porto, *Science* **167**, 1715 (1970); S. L. Kurtin, C. A. Mead, W. A. Mueller, B. C. Kurtin, and E. D. Wolf, *Science* **167**, 1720 (1970); S. W. Rabideau and A. E. Florin, *Science*, **169**, 52 (1970). These papers report the presence of significant quantities of Na, K, Ca, B, Cl, and SO₄ in material alleged to be "polywater." At press time, the existence of polywater remains highly controversial.

extremely useful, is arbitrary and restrictive. Hydrogen peroxide does have one weak bond per molecule, the O—O energy being about 34 kpm. Nevertheless it is thermodynamically stable in the usual sense as indicated by the negative enthalpy of formation. It is explosively unstable, however, because it is *not* thermodynamically stable with respect to other products, only one of which happens to be a free element. Such compounds are common. In addition, as we shall see, numerous compounds that do not normally exist in a stable state are perfectly possible in the sense of being held together by strong bonds. They are unknown or almost unknown because of the ease with which they can be transformed into still more stable substances.

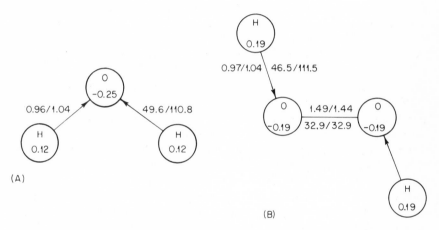

FIG. 6-2. Schematic representation of (A) water and (B) hydrogen peroxide.

Why Carbon Dioxide Doesn't Polymerize

The polymerization of olefins is an exothermic process occurring readily in the presence of appropriate catalysts and conditions. Then why doesn't carbon dioxide, with presumably similar double bonds, also polymerize?

The atomization energy of carbon dioxide, CO_2 gas, is determined in the usual way, treating each bond as a double bond like that in ethylene or oxygen. The partial charges on carbon and oxygen are found to be 0.22 and -0.11, giving an ionic weighting coefficient t_i of 0.17. The geometric mean of the O'' energy 66.7 and the carbon energy 83.2 is 74.5. The bond length is 1.16 Å compared to the covalent radius sum of 1.49 Å. The covalent energy contribution is then

$$t_c E_c = \frac{0.83 \times 74.5 \times 1.50 \times 1.49}{1.16} = 119.1 \text{ kpm}$$

The ionic energy contribution is

$$t_i E_i = \frac{0.17 \times 332 \times 1.50}{1.16} = 73.0 \text{ kpm}$$

The sum, 192.1, multiplied by 2 equals 384.2, the calculated atomization energy. The experimental value is accurately known to be 384.6 kpm. The standard heat of formation is -93.7 calculated and -94.1 experimental.

One reason for giving these calculations in full detail is that the bonds in carbon dioxide have long been recognized as appreciably stronger, by 15–20 kpm, than average carbon–oxygen double bonds in aldehydes, ketones, and other organic compounds. They are also significantly shorter, 1.16 compared to an average 1.21 Å. These facts have prompted the suggestion that some single–triple bond resonance is involved in CO_2 to cause the bonds to be abnormally strong. The calculations indicate such an assumption to be unnecessary. The bonds in CO_2 are quantitatively described as ordinary polar double bonds. They are stronger than the average carbonyl bond in organic compounds because they are shorter and because the ionic energy coefficient is a little larger, 0.17 instead of 0.15. The latter difference results in part from the fact that the partial charge on carbon in CO_2 is much more positive than in compounds in which the positive charge is shared by other atoms. The higher positive charge on carbon makes it smaller and the bond shorter.

A carbon dioxide polymer would be assumed to involve four carbon–oxygen bonds per carbon atom, each single and therefore involving the O' parameter of 34.0 instead of the O'' parameter of 66.7. The geometric mean covalent energy with carbon would then be 53.2 and the 1.50 multiplicity factor would no longer apply. The bonds would be longer, about 1.45 Å compared to the covalent radius sum of 1.49 Å. The covalent contribution would then be

$$t_c E_c = \frac{0.83 \times 53.2 \times 1.49}{1.45} = 45.4 \text{ kpm}$$

The ionic contribution would be

$$t_i E_i = \frac{0.17 \times 332}{1.45} = 38.9 \text{ kpm}$$

The sum, 84.3, is the energy of a single C—O bond in the polymer. Atomization would require breaking four such bonds per carbon atom, the energy being 337.2 kpm (Fig. 6-3). This is less than that of molecular CO_2 by 47.4 kpm. In other words, the enthalpy of polymerization of CO_2 gas would be $+47.4$ kpm. The entropy would decrease in the polymerization process, leading to a free energy of polymerization in the neighborhood of $+50$ kpm.

It is easy to understand from the above calculations why CO_2 remains monomeric. The difference in bond energy parameters of oxygen causes this preference. Suppose the same O'' parameter were applicable to the singly bonded polymer. Then the covalent contribution would be 63.6, giving a total bond energy of 102.5 kpm. Multiplied by 4 this would become 410.0 kpm, 25 kcal larger than the atomization energy of the monomeric gas. If this were so, carbon dioxide would be a solid like SiO_2.

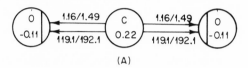

(A)

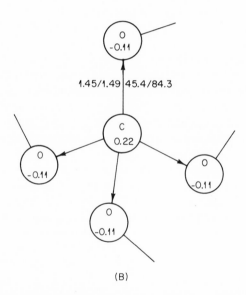

(B)

FIG. 6-3. Schematic representation of carbon dioxide. (A) Monomer, atomization energy: 384.2 kpm, calc.; 384.6 kpm, exp. (B) Polymer, atomization energy: 337.2 kpm, calc.; heat of polymerization: +47.4 kpm.

Why Silicon Dioxide Is Solid

One of the best known anomalies of the periodic table is of course the monomeric nature of carbon dioxide, but it would not be anomalous if it were not for the polymeric nature of the other oxides of the group. An examination of the nature of silicon dioxide is therefore in order.

Silicon dioxide exists normally as a high melting solid polymer in which each silicon atom is surrounded by four oxygen atoms, each of which bridges two silicons. The angle of the oxygen bridge is about 150°, suggesting a considerable degree of involvement of the lone pairs on oxygen. But SiO_2 does boil, at about 2600°C, and SiO_2 molecules exist in the vapor. Let us first consider the monomer, since energy calculations disclose a curious difference from carbon dioxide.

The atomization of $SiO_2(g)$ experimentally determined is 304.3 kpm. The bond length is 1.54 Å compared to the covalent radius sum of 1.89 Å. If the bonds were ordinary double bonds, their energy would be calculable using the Si bond energy parameter of 53.4 and the O'' parameter of 66.7. The geometric mean energy is 59.7. The partial charges are 0.40 on silicon and -0.20 on oxygen giving a value of 0.30 for t_i. The covalent contribution to the double bond would be

$$t_c E_c = \frac{0.70 \times 59.7 \times 1.89 \times 1.50}{1.54} = 76.9 \text{ kpm}$$

The ionic contribution would be

$$t_i E_i = \frac{0.30 \times 332 \times 1.50}{1.54} = 97.0 \text{ kpm}$$

The sum, 173.9, is the calculated double bond energy. However, twice this value is far greater than the experimental atomization energy of 304.3 kpm. A different model seems necessary.

Where vacant orbitals are available on the other atom, oxygen commonly acts as acceptor in coordinating to an electron pair, making use of its O''' bond energy parameter. Numerous examples of this have been observed, such as most of the oxyhalides of phosphorus and sulfur. Suppose that for some reason the outer shell structure of silicon should remain s^2pp instead of sp^3. Then one oxygen could form a double bond using the silicon p orbitals, but the other would coordinate to the s^2 pair. The two bonds would of course become exactly alike through averaging or hybridization of the bonds. The O''' parameter is 99.4, giving a geometric mean of 72.8 with silicon. The covalent contribution to a Si—O''' bond would be

$$t_c E_c = \frac{0.70 \times 72.8 \times 1.89}{1.54} = 62.5 \text{ kpm}$$

The ionic contribution would be

$$t_i E_i = \frac{0.30 \times 332}{1.54} = 64.7 \text{ kpm}$$

The sum is the bond energy, 127.2 kpm. This energy added to the double bond energy of 173.9 gives the total atomization energy of 301.1, reasonably close to the experimental value of 304.2 kpm. A tentative conclusion is that gaseous SiO_2 does not have two ordinary double bonds per molecule, like CO_2. Instead, it averages one double and one coordinate covalent bond. As will be discussed a little later, this is by no means unique in chemical bonding. The same kind of bond has been observed in SO and SO_2 and perhaps also in the carbonyl halides.

In the solid, there exists a network of covalent bonds. The energy of these can be calculated just as easily as if they were part of a small molecule instead of a giant polymeric structure. The bond lengths are reported to vary from 1.52 to 1.69 Å, causing some doubt as to choice. The average, 1.61 Å, is selected. The bonds may be described as single Si—O″ bonds, using the O″ parameter of 66.7. The geometric mean covalent energy is 59.7. The covalent contribution is

$$t_c E_c = \frac{0.70 \times 59.7 \times 1.89}{1.61} = 49.1 \text{ kpm}$$

The ionic contribution is

$$t_i E_i = \frac{0.30 \times 332}{1.61} = 61.8 \text{ kpm}$$

The sum is 110.9 kpm, the Si—O″ energy. Atomization requires breaking four bonds per silicon atom or an energy of 443.6 kpm. The experimental atomization energy is in excellent agreement, 445.6 kpm, strongly supporting the assignment of bond type.

It is of course perfectly clear from the standard heats of formation of SiO_2(g) – 76.2 and SiO_2(c) – 217.7 that polymerization of the gas is highly exothermic (see Fig. 6-4). It is interesting to examine the cause. Even if SiO_2(g) did contain two double bonds per molecule, which would make its standard heat of formation about – 120 kpm, the polymerization would still be exothermic by nearly – 100 kpm. Here the greater stability of two single bonds compared to one double bond, inevitable when the same O″ parameter is applicable to both single and double bond, becomes a very dominant factor due to the much higher polarity of the silicon–oxygen bond. The apparent fact that silicon does not simultaneously form two double bonds to oxygen in the gaseous SiO_2 molecule causes the advantage of polymerization to be even greater.

Why Carbon Disulfide Is Monomeric

It is interesting that when sulfur acquires a partial negative charge, it appears to form double bonds in which the S‴ parameter is applicable. This is true in both CS_2 and COS. For example, in CS_2 the partial charges

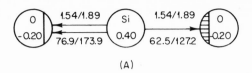

(A)

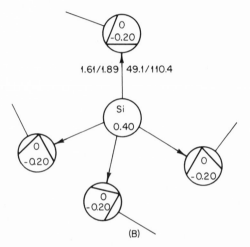

(B)

FIG. 6-4. Schematic representation of silicon dioxide. (A) Monomer (both bonds are equal); atomization energy: 301.1 kpm, calc.; 304.3 kpm, exp. (B) Polymer, atomization energy: 443.6 kpm, calc.; 445.8 kpm, exp.

are 0.05 on carbon and -0.03 on sulfur, leading to a value of 0.04 for t_i. The S''' parameter is 69.0, giving a geometric mean with carbon of 75.8. The bond length is 1.56 Å and the covalent radius sum is 1.81 Å. For a double bond of type $C=S'''$, the covalent contribution is

$$t_cE_c = \frac{0.96 \times 75.8 \times 1.50 \times 1.81}{1.56} = 126.6 \text{ kpm}$$

The ionic contribution is

$$t_iE_i = \frac{0.04 \times 332 \times 1.50}{1.56} = 12.8 \text{ kpm}$$

The sum, 139.4, times two bonds equals 278.8, the calculated atomization energy of $CS_2(g)$. This is in good agreement with the experimental value of 276.4.

The question of why CS_2 is monomeric is intriguing because the usual bond energy approach reveals that it ought to be polymeric. Here the low polarity gives the covalent bonding energy its maximum effect. In sulfur, however, the advantage of multiplicity compared to oxygen is greatly diminished, since the difference between single prime and double or here even triple prime energies is not nearly as great. The S' energy of 55.0 gives a geometric mean with carbon of 67.7. A single bond of such low polarity would have a length close to the covalent radius sum. In fact, this is shown in the experimental bond length in mercaptans and thioethers. A bond length of 1.81 is therefore assigned to the hypothetical polymer. The covalent energy contribution is then

$$t_c E_c = \frac{0.96 \times 67.7 \times 1.81}{1.81} = 65.0 \text{ kpm}$$

The ionic energy is

$$t_i E_i = \frac{0.04 \times 332}{1.81} = 7.3 \text{ kpm}$$

The sum, 72.3, is the C—S bond energy in the polymer. Atomization would require breaking four such bonds per carbon atom, or an energy of 289.2 kpm. Thus it appears that the heat of polymerization should be $276.4 - 289.2 = -12.8$ kpm. Then why is carbon disulfide a monomeric liquid instead of a polymeric solid?

The data for carbon tetrachloride provide an answer. The calculated atomization energy of $CCl_4(g)$ is 35 kpm higher than the experimental value. A detailed explanation is given in the following chapter. Briefly, the 35 kcal represents a reduction in the total bonding energy caused by the fact that attachment of four chlorine atoms to the same carbon atom requires mutual distortion of their electronic spheres. The same steric requirement would hold for sulfur atoms. They are nearly the same size as chlorine atoms. The environment of a carbon atom in polymeric CS_2 would be very similar to its environment in CCl_4, closely surrounded by large atoms so packed together that they crowd one another. The repulsion energy would presumably be very similar. The net enthalpy of polymerization of CS_2 would therefore be more like $+22$ kpm than -13. So CS_2 is not polymeric after all.

CARBONYL SULFIDE

The molecule of carbon oxysulfide, or carbonyl sulfide, COS, is halfway between CO_2 and CS_2. Presumably the crowding in a polymer in which each carbon atom is surrounded by two sulfur atoms and two oxygen atoms would be negligible and could not be taken as an excuse for failure to polymerize. Bond energy calculations clearly reveal why it remains monomeric. For this molecule, the energy of the C=O″ bond is calculated to be 189.3 kpm. The energy of the C=S‴ bond is calculated to be 139.4. Their sum, 328.7, compares favorably with the experimental atomization energy of 331.5 kpm.

In the polymer, the O' and S' parameters would be applicable and the single bonds would be longer than the double bonds of the monomer. A C—O single bond energy is calculated as 81.8 and a C—S single bond energy as 72.4 kpm. Two of each would be broken in atomizing the polymer, corresponding to an atomization energy of 308.4 kpm. The difference between this value and the experimental atomization energy of the monomer is 23.1. In other words, polymerization would be endothermic by 23 kpm. The decrease in entropy that would accompany polymerization would ensure a high positive free energy change for this reaction. No wonder COS remains monomeric.

Carbon Monoxide and Its Analogs

Until the dissociation energy of carbon monoxide could be measured unambiguously, the bond in the nitrogen molecule N_2 with its dissociation energy of 226 kpm was believed to be the strongest of known bonds. The discovery that the bond in CO gas is stronger yet arouses some curiosity as to why. The molecule is of course isoelectronic with N_2. The bond is evidently one in which the carbon atom only contributes two of its valence electrons but provides a vacant orbital for an otherwise unused lone pair of the oxygen. In other words, the triple bond could be described as consisting of two normal covalent bonds and one coordinate covalent bond, with oxygen donor. Nevertheless, the usual calculation of partial charge appears quite satisfactory.

Electronegativity equalization leads to a partial charge of 0.16 on carbon and -0.16 on oxygen. Thus t_i is 0.16. The bond length is only 1.13 Å, whereas the covalent radius sum is 1.49 Å. The O''' parameter of 99.4 gives a geometric mean with carbon of 91.0. The multiplicity factor for a triple bond is 1.77. The covalent energy contribution is then

$$t_c E_c = \frac{0.84 \times 1.77 \times 91.0 \times 1.49}{1.13} = 178.4 \text{ kpm}$$

The ionic energy contribution is

$$t_i E_i = \frac{0.16 \times 332 \times 1.77}{1.13} = 83.2 \text{ kpm}$$

The total bond energy, 261.6 kpm, agrees reasonably well with the experimental value of 257.3 kpm. Since the heteronuclear covalent energy parameter is a little smaller in CO (91.0) than the equivalent parameter for N_2 (94.8), the principal advantage possessed by the CO molecule is its polarity. Carbon monoxide has a stronger bond than the nitrogen molecule because of the large contribution of ionic energy in carbon monoxide.

A molecule that seems closely analogous to CO is CS, which has been found to have a dissociation energy of 166 kpm. Its bond length is 1.53 Å. The covalent radius sum is 1.81 Å. The geometric mean homonuclear bond energy parameter for C—S''' is the same as used for CS_2 calculation, 75.8. The covalent energy contribution is

$$t_c E_c = \frac{0.96 \times 75.8 \times 1.77 \times 1.81}{1.53} = 152.4 \text{ kpm}$$

The ionic contribution is

$$t_i E_i = \frac{0.04 \times 332 \times 1.77}{1.53} = 15.4 \text{ kpm}$$

The sum, 167.8, agrees well with the experimental bond energy of 166, showing that the molecule has a triple bond as assumed.

One might accordingly expect silicon monoxide gas, SiO, also to be analogous to carbon monoxide. If it did have a triple bond, the energy would be calculated to be 227.5 corresponding to a standard heat of formation of −59.0. However, the standard heat of formation selected by the

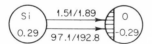

Fig. 6-5 Schematic representation of silicon monoxide. Atomization energy: 192.8 kpm aclc.; 192.3 kpm, exp.

National Bureau of Standards in 1966 is only −23.8 kpm. This corresponds to an atomization energy of 192.3 kpm. It is therefore interesting that if the O''' parameter is used, but for some unknown reason perhaps associated with the relative sizes of the two atoms, oxygen is unable to form a triple bond to silicon, it might form a double bond instead: Si=O'''. The atomization energy for this bond is calculated to be 192.8 kpm (see Fig. 6-5). This would correspond to a standard heat of formation of −24.3 kpm, in close agreement with the accepted value. We may conclude tentatively then, that silicon is unable to form a triple bond to oxygen and that the double bond assumption is correct. This may prove consistent when better understanding is reached with the observation previously discussed—that when an atom of silicon has formed one double bond to one oxygen atom it appears unable to form another double bond to a second oxygen atom.

Both carbon monoxide and silicon monoxide are unstable with respect to the dioxide and carbon or silicon. The enthalpy of disproportionation of carbon monoxide to carbon and carbon dioxide is -41.3 kcal, or -20.7 kpm of CO. The value for SiO is -170.1 or -85 kpm of SiO. Yet at ordinary temperatures carbon monoxide shows no tendency whatever to undergo this change, whereas silicon monoxide appears to change rapidly when cooled to a very finely divided mixture of silicon and silicon dioxide. This great difference between CO and SiO is probably related to this bonding difference as well as to the high stability of $SiO_2(c)$. A triple bonded SiO would still tend to disproportionate but might be expected to be kinetically more stable.

Interestingly, SiS does appear to contain a triple bond, analogous to that of CO, supporting the suggestion that disparity of size may prevent a triple bond in SiO. The partial charges in SiS are 0.17 and -0.17. The S''' parameter of 69.0 gives a geometric mean with 53.4 for silicon of 60.8. The covalent radius sum is 2.21, compared to the observed bond length of 1.93. The covalent energy contribution is

$$t_c E_c = \frac{0.83 \times 60.8 \times 1.77 \times 2.21}{1.93} = 102.3 \text{ kpm}$$

The ionic contribution is

$$t_i E_i = \frac{0.17 \times 332 \times 1.77}{1.93} = 51.8 \text{ kpm}$$

The sum is the atomization energy, 154.1 kpm. The experimental value is 158.6 kpm.

The Instability of Carbonic Acid

Salts of oxyanions have been observed generally to tend toward greater thermal stability, the more completely the "oxyanions" control the bonding electrons.[3] Protons, having great polarizing power, tend to destroy the symmetry of the oxyanion groups by attracting electrons to the hydrogen. Thus the stability is reduced. It is common knowledge that metal salts, particularly of the less electronegative metals, tend to be much more stable than the free oxyacids, some of which cannot be isolated at all.

Additional insight as to the instability of carbonic acid can be gained by calculating the bond energy of the gaseous molecule. One may reasonably expect a certain degree of stabilization of a compound through its solvation, especially in water, if, like H_2CO_3, it is capable of protonic bridging. Carbonic acid molecules can perhaps exist in small concentrations in aqueous solution. The heat of formation of the gas can be calculated to learn why it is unknown out of water.

[3] A. G. Ostroff and R. T. Sanderson, *J. Inorg. Nucl. Chem.* **4,** 230 (1957).

The gaseous compound being unknown, bond lengths must be estimated on the basis of experimental values for similar molecules. The individual bond lengths so assigned are C—O, 1.38; O—H, 0.95; and C=O, 1.25. As will be shown in a later chapter, when one oxygen atom is singly bonded to a carbonyl carbon, its bond energy reflects the O'' parameter instead of the O' parameter. However, when more than one oxygen atom is singly bonded to a carbonyl carbon, this is not true. Then the O' parameter is applicable to all but the doubly bonded oxygen. By the usual methods, the bond energies are calculated as OH, 113.7; C—O, 86.7; C=O, 175.6. The atomization energy of $H_2CO_3(g)$ is 576.4, which corresponds to a standard heat of formation of -122.1 kpm. In other words, with respect to its free elements, the carbonic acid molecule should be very stable.

However, as will be considered in more detail in a later chapter, there is abundant evidence that the presence of two or more adjacent hydroxyl groups on the same carbon presents a very favorable opportunity for the splitting off of water. Superficially this might appear to offer little advantage since, as calculated above, two C—O bonds are very nearly the equivalent of one C=O bond in this molecule and the decomposition of H_2CO_3 to CO_2 and H_2O only changes two C—O bonds to one C=O bond. However, here the advantage lies in the fact that each C=O bond in CO_2 is substantially stronger (by 15 kpm or more) than ordinary carbonyl bonds. The standard heat of formation of $H_2O(g)$ is -57.8 and that of $CO_2(g)$ -94.1 giving a total of -151.9 comparable to only -122.1 for H_2CO_3. Thus the decomposition is exothermic by -29.8 kpm. Together with the kinetic ease of water splitting provided by the structure of this molecule, this accounts very well for the nonexistence of carbonic acid molecules in the gaseous state.

A similar calculation can be done to explain the nonexistence of carbon tetrahydroxide, or orthocarbonic acid, $C(OH)_4$. On paper this compound appears to be just as reasonable as, for example, CF_4, which is a well-known compound.

Oxides of Nitrogen

As a class the nitrogen oxides are volatile substances containing multiple bonds showing little or no tendency to polymerize. As a class they are very different from the oxides of the heavier elements of this group. These differences may be accounted for quantitatively by the advantages of bond multiplicity especially when, as in nitrogen oxides, bond polarity is low. Bond energy calculations can be very useful in aiding understanding of the bonding in these compounds, for they sometimes disclose bonding situations different from conventional concepts.

Nitrous oxide, N_2O, is an example of a molecule in which the bonding must be unusually complex. It is known to be linear, with a sequence represented by NNO. It would seem reasonable for the terminal nitrogen to be attached to the central nitrogen by a triple bond except that then the oxygen would have no point of attachment but the nitrogen lone pair. Even though molecular nitrogen is known to function as ligand in a few complexes of the transitional metals it seems an unlikely donor toward oxygen. Moreover, the atomization energy would be too high even if the single bond were N′—O′. Let us assume that the oxygen bonds normally by a double bond to the central nitrogen. This would leave the central nitrogen incapable of forming a triple bond to the terminal nitrogen but still possessing three otherwise unoccupied electrons in the valence shell. If the NN bond is represented as N‴—N‴, using no multiplicity factor, its energy is calculated as 118.0 kpm. The energy of the N″=O″ bond is calculated as 146.6 kpm. Their sum, 264.6, is in close agreement with the experimental atomization energy of 266.1 kpm. The ease with which N_2O serves as a source of oxygen, for example in promoting combustion, suggests the ease with which the bonding capacity of the central nitrogen transfers from the bond to oxygen to the bond to the other nitrogen. For purposes of bond energy calculation, nitrous oxide appears to be quite adequately represented as N‴—‴N″=O″.

Nitric oxide, paramagnetic by virtue of its odd number of electrons, is usually described as having a bond order of 2.5, the double bond being augmented by an extra electron from the nitrogen. However, bond energy calculations offer no suggestion of such additional bond strength. A normal N″=O″ bond is calculated to have a covalent energy of 117.0 and an ionic energy of 34.6, giving a total atomization energy of 151.6. The experimental value is 151.0 kpm. Here, incidentally, is an example of a thermodynamically unstable molecule having a very strong bond. The standard heat of formation is calculated to be +21.0 kpm, and measured as +21.6. Nevertheless there seems to be no easy mechanism for reversion to the elements. Nitric oxide retains its identity over a wide temperature range.

The presence of the single outer electron on the molecule suggests a possible ease of dimerization which is not in fact observed. Placing this electron in the NO bond removes it at least partially from this possibility, but would presumably enhance the bond strength for which the present calculations give no evidence. Evidently, however, the electron is so placed that it is not readily available for dimerization. The molecule is readily oxidized, nevertheless, forming nitrogen(IV) oxide, NO_2, which dimerizes easily at lower temperatures.

NO_2 can be interpreted as averaging one N′—O″ bond and one N″=O″ bond. The calculated atomization energy of this structure is 226.7 kpm compared to an experimental value of 224.2 kpm (see Fig. 6-6).

N_2O_5 is a difficult molecule to interpret. Tentatively, it may be described as averaging three N''—O' and three N''—O'' bonds. This gives a total atomization energy of 522.9 compared to 521.0 for the experimental value.

Nitrogen Oxyacids

Another seemingly reasonable compound that is unknown is nitrogen trihydroxide, or orthonitrous acid. Assuming the bond lengths to be 0.96 for O—H and 1.44 for N—O, one can calculate the atomization energy and thus the standard heat of formation of $N(OH)_3(g)$ to be -43.6 kpm. It would

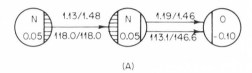

(A)

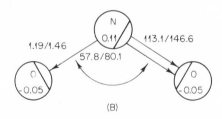

(B)

FIG. 6-6. Schematic representation of (A) nitrous oxide and (B) nitrogen dioxide (both bonds are equal). (A) Atomization energy: 264.6 kpm, calc.; 266.1 kpm, exp. (B) Atomization energy: 226.7 kpm, calc.; 224.7 kpm, exp.

therefore be a thermodynamically stable substance in the conventional sense. However, the juxtaposition of hydroxyl groups on the same nitrogen make water splitting relatively simple. The standard heats of formation of nitrous acid, HNO_2, and H_2O are -27.3 and -57.8, totaling -85.1. The calculated enthalpy of the decomposition of nitrogen trihydroxide to nitrous acid and water is -41.5 kpm. Together with the kinetic ease of dehydration, this exothermicity explains the nonexistence of orthonitrous acid.

Nitrous acid itself is not stable, but its standard heat of formation in the gas state is reported as -18.9 kpm. Calculation based on the assumption that the bonds are O—H, N'—O', and N''—O'' gives a somewhat higher value of -27.0 kpm. Whatever the correct value, if this is of the right order of magnitude, the familiar instability of this compound must be related to the probable ease with which its atoms rearrange to form stronger bonds in different compounds.

The two lone oxygens on the molecule of nitric acid, HNO_3, are evidently exactly equivalent yet the bonds can be represented as though on the average one were N'—O' and the other N''=O''. The bond to hydroxyl oxygen is of course N'—O' but longer since no multiplicity is involved. The calculated atomization energy based on this model is 373.5 kpm, in very satisfactory agreement with the experimental value of 376.2 kpm (see Fig. 6-7). In

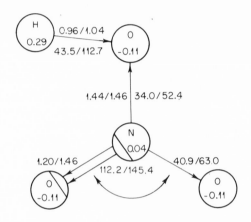

FIG. 6-7. Schematic representation of nitric acid. Atomization energy: 373.5 kpm, calc.; 376.2 kpm, exp.

making the calculations no account was taken of the experimental evidence that restricted rotation about the hydroxyl bond to nitrogen suggests some internal protonic bridging. However, this would hardly add more to the calculated value than the difference noted above.

Since both ortho- and metaphosphoric acids are known, the question may well be raised as to why there is no orthonitric acid, H_3NO_4. This molecule would of necessity consist of three normal single bonds between nitrogen and hydroxyl oxygen plus one coordinate covalent bond between nitrogen donor and oxygen acceptor. On this basis, and with the assignment of reasonable bond lengths, one can calculate the heat of formation of ortho-nitric acid in the gas state to be -42.8 kpm. However, the sum of the standard heats of formation of HNO_3 and H_2O is -90.1, showing that the enthalpy of dehydration of gaseous orthonitric acid would be -47.3 kpm. Together with the probable ease of water splitting from a grouping of three hydroxyl groups attached to the same nitrogen atom, this explains the nonexistence of orthonitric acid, H_3NO_4.

One might also ask the question: Why isn't HNO_3 polymeric like HPO_3?

To polymerize, nitric acid would in effect need to substitute two N'—O' single bonds of 52.4 kpm energy for the short N'—O' bond of 63.0 energy and the double bond of 145.4 kpm energy. In other words, bonds totaling 105.0 kpm energy in the polymer would be substituting for bonds totaling 208.4 kpm in the monomer, causing the enthalpy of polymerization to be about + 100 kpm. Here the reduction in lone pair weakening that accompanies bond multiplicity offers a large advantage to the monomer over the polymer.

Oxides of Phosphorus

Phosphorus is much less electronegative than nitrogen. Consequently its bonds to oxygen are much more polar, and the polar contribution to the bonding energy is larger. The advantages to multiple bonding shown by nitrogen in its oxides are also much less with phosphorus, since the difference in energy between P' and P''' is very much smaller. Phosphorus does have outer d orbitals not possessed by nitrogen, however, which appear to provide a means of reducing the bond weakening by the lone pairs of oxygen.

The gas molecule PO is reported to have a dissociation energy of only 125.0 kpm. This is considerably less than would be expected of a double bond. If one assumes, however, a P'''—O''' bond, the energy is calculated as 125.3 kpm, in almost exact agreement with the experimental value. The implication of this bond type is that the oxygen is acting as an acceptor and the phosphorus as a donor, under such circumstances that the lone pair weakening is removed in each atom. Obviously this isolated fact hardly justifies extensive speculation, but it clearly calls for further investigation. One may compare, for example, the disparity of size and reduced multiplicity in Si=O''' and P'''—O'''. The subject arises again in Chapter 8.

Unless we can be sure that in the more conventional oxides, phosphorus exhibits only its P' energy, the calculation method offers too many alternatives to serve usefully in the identification of bond type. In practically all compounds studied, however, unless normal double or triple bonds are formed, the positively charged atom of the single bond uses only its E' parameter. The negatively charged atom may then exhibit E', E'', or E''' parameters. If this is applied to phosphorus oxides, then the best representation of $P_4O_6(g)$ seems to be that its bonds are an average of P'—O'' and P'—O''' single bonds. This model leads to a total atomization energy of 1188.0 kpm, compared to the experimental value of 1187.5.

The representation of $P_4O_{10}(g)$ is even more speculative. Its oxygen bridge bonds appear to be an average of P'—O' single bonds and P'—O'' single bonds. Good agreement with the experimental atomization energy then requires that the four terminal oxygen bonds at the corners of the molecule must be P'''—O''' single bonds. It may be pointed out that, in general, oxygen in oxyhalides of both phosphorus and sulfur is attached in coordinate covalent bonds of the type E'—O'''. With phosphorus, the bond lengths in the oxyhalides are about 1.45 Å, consistently, whereas in P_4O_{10}, the bond lengths to the terminal oxygens are only 1.39 Å. A better understanding of the nature of such bonds is certainly needed. The total calculated atomization energy for such a model is 1596.0 kpm, whereas the experimental value is 1588.7 kpm.

The bonds in $As_4O_6(g)$ seem adequately described as As'—O'' single bonds. The bond energy is calculated to be 77.9 kpm, and there are twelve bonds per molecule. The total atomization energy is then 934.8 kpm, to be compared with the experimental value of 935.8 kpm. However, agreement with this model might prove fortuitous, since difficulties are encountered with other arsenic compounds, notably the trihalides.

Oxides and Oxyacids of Sulfur

One of the notable curiosities about sulfur–oxygen chemistry is the absence among well-known compounds of the monoxide, SO. For if O_2 and S_2 are both very stable molecules, why would not SO, which has the advantage of polarity, be even more stable and a familiar substance?

Let us prepare for a discussion of SO by first considering the common product of burning sulfur, SO_2. This compound consists of molecules, SO_2, in which the presence of one lone pair on the sulfur that is not involved in the bonding is suggested by the observed bond angle of about 120°. The bond length is 1.43 Å, the covalent radius sum 1.76. The partial charge on sulfur is 0.16 and on oxygen -0.08, giving an ionic weighting coefficient of 0.12. If the bonds are assumed to be ordinary double bonds, S''=O'', then the double prime parameters 62.1 for sulfur and 66.7 for oxygen are applicable. Their geometric mean is 64.4. The multiplicity factor is 1.50 for a double bond. The calculated covalent energy contribution is

$$t_cE_c = \frac{0.88 \times 64.4 \times 1.76 \times 1.50}{1.43} = 104.6 \text{ kpm}$$

The ionic contribution is

$$t_iE_i = \frac{0.12 \times 332 \times 1.50}{1.43} = 41.8 \text{ kpm}$$

The sum, 146.4 kpm, is the energy of a sulfur–oxygen double bond. Two such bonds would give an atomization energy of SO_2 of 292.8 kpm. But the experimental value is only 256.7 kpm. What is the origin of this discrepancy?

Let us suppose that only the two "normal" bonding orbitals of the sulfur atom are available for forming an ordinary double bond. The second bond would then be a coordinate covalent bond of type S'—O'''. For such a bond, the geometric mean of the sulfur parameter 55.0 and the oxygen parameter of 99.4 is 73.9. The covalent energy contribution is

$$t_cE_c = \frac{0.88 \times 73.9 \times 1.76}{1.43} = 80.0 \text{ kpm}$$

The ionic contribution is

$$t_iE_i = \frac{0.12 \times 332}{1.43} = 27.9 \text{ kpm}$$

The sum, 107.9, is the S'—O''' bond energy in SO_2. Addition of the calculated double bond energy of 146.4 kpm to this single bond energy gives the sum 254.3, in excellent agreement with the experimental atomization energy of 256.7 kpm. The individual bonds in the SO_2 molecule according to this evidence may be accurately represented as the average of an ordinary double bond and a coordinate covalent bond. This kind of bonding was previously observed, it may be recalled, in gaseous SiO_2.

We may now return to the SO molecule (see Fig. 6-8). Its bond length has been reported to be 1.49 Å, a little longer than in SO_2. Its atomization energy has been experimentally determined as 124.7 kpm. The partial charges on sulfur and oxygen are 0.12 and −0.12. From these data one can calculate that if the bond were a double bond, its energy would be 140.5. This is substantially higher than that of either O_2 (119.2) or S_2 (102.4)

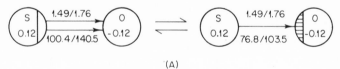

(A)

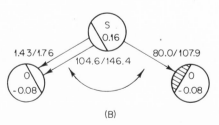

(B)

FIG. 6-8. Schematic representation of SO and SO_2. (A) Sulfur monoxide (bond is an average of the two); atomization energy: 122.0 kpm, calc.; 124.7 kpm, exp. (B) Sulfur dioxide (bonds are equal); atomization energy; 254.3 kpm, calc.; 256.7 kpm, exp.

because of the advantage of the bond polarity. But it is also higher than the experimental energy. On the other hand, if the bond were represented as S'—O''', the energy would be calculated to be 103.5 kpm. The average of these two kinds of bonds, 122.0 kpm, agrees well with the experimental value of 124.7 kpm. It is concluded that the bond in the SO molecule is essentially like a bond in SO_2—an average of a double bond and a co-ordinate covalent bond. This still makes it more stable than either the O_2 or S_2 molecule, so why isn't it a well-known species?

The answer must lie in the ease with which SO disproportionates to sulfur and SO_2. This process apparently occurs extremely readily. The standard heat of formation of SO is +1.5 kpm, whereas that of SO_2 is −70.9. For the reaction,

$$2SO(g) \rightarrow S(c) + SO_2(g)$$

where $\Delta H = -73.9$ kpm, entropy would decrease in the conversion of 2 moles of gas to 1 mole of gas and 1 of crystalline solid, but the free energy for this disproportionation would still have a high negative value. Sulfur monoxide is an example of many little-known or unknown compounds which would have amply strong chemical bonds and often high negative heat of formation. Neither stability against reversion to the free elements nor strong bonds ensure stability against some other kind of atomic rearrangement which could combine kinetic ease with exothermicity. Other examples, such as nitrogen trihydroxide, have been mentioned, and others will be encountered.

Indeed, a closely related example may now be considered. For many years, what was interpreted as SO appears to have been actually S_2O, a molecule quite unexpected because of its stoichiometry. This compound has recently been studied and found to have a bent structure resembling SO_2, but with one sulfur in place of one of the oxygen atoms. It is interesting that bond energy calculations support the suggestion that equality of size helps in multiple bond formation. The double bond appears to occur between the two sulfur atoms. The oxygen atom is attached by a coordinate covalent bond. Calculated in the usual way, the energy of the S''=S'' bond is 94.8 kpm and that of the S'—O''' bond 104.9. From the total, 199.7 kpm, the standard heat of formation is calculated to be −6.9 kpm. Again disproportionation occurs evidently very readily: $2\,S_2O(g) \rightarrow 3\,S(c) + SO_2(g)$; $\Delta H = -57.1$ kpm. So another potentially stable compound has only a transitory existence because of the ease with which it changes to more stable products.

Returning now to SO_2, we face one of the difficult questions of general chemistry that has never been answered satisfactorily. Why is the combination of SO_2 with water so very different from that of SO_3 with water? Why is sulfuric acid a stable compound whereas sulfurous acid has never been isolated?

To answer this question the atomization energy and heat of formation of gaseous sulfurous acid, H_2SO_3, are first needed. Some reasonable assumption about the nature of the bonds is needed. Instead of viewing the bonds in SO_2 in their true light of representing an average of a double and a single bond, let us consider a hypothetical static situation in which one bond is actually double, $S''{=}O''$, and the other bond single, $S'{-}O'''$. The natural place for water to add would be at the double bond. This would produce two $S'{-}O'H$ bonds in place of the double bond. The third sulfur to oxygen bond would remain unchanged. Consistent with this concept is the evidence, also from bond energy calculations, that thionyl fluoride and chloride consist of sulfur attached to two halogen atoms and to an oxygen through an $S'{-}O'''$ bond. If this kind of an S—O bond occurs in OSX_2, then why not in $OS(OH)_2$?

Bond lengths assumed for the H_2SO_3 gas molecule are those reported for $H_2SO_4(g)$: O—H, 0.97; S—OH, 1.53; and S—O''', 1.42. The calculated atomization energy of such a molecule is 475.5 kpm, corresponding to a standard heat of formation of -125.9. From the viewpoint of bond strength, H_2SO_3 should be a stable molecule. However, two hydroxyl groups attached to the same atom of sulfur suggest an ease of dehydration. The heats of formation of water and SO_2, both gaseous, total -128.7, corresponding to an enthalpy of dehydration of -2.8. In the gas phase this reaction would be favored by an increase in entropy, giving a negative free energy change in the neighborhood of -7 kpm. If both sulfurous acid and water were in the liquid phase, their heats of formation would be increased approximately equally by protonic bridging, but then the entropy would still be increased by the release of sulfur dioxide gas. The data thus account reasonably well for the instability of H_2SO_3 molecule.

The situation with respect to SO_3 and water is quite different. Here the gas phase standard heats of formation are H_2SO_4, -177.0; SO_3, -94.6; and H_2O, -57.8. The enthalpy of combination of SO_3 and H_2O in the gas phase is -21.0 kpm. Although the entropy would decrease somewhat, the free energy change would still be substantially negative, ensuring the stability of $H_2SO_4(g)$ with respect to dehydration. These are all experimental values. It is of interest to consider the calculations. The results of these calculations are summarized in Table 6-1, where the effects of hydration of SO_2 and SO_3 can be compared on a bond-for-bond basis.

Like SO_2, SO_3 averages in only one true double bond per molecule (see Fig. 6-9). This is averaged with two single bonds that can best be described as $S''{-}O''$ bonds instead of the $S'{-}O'''$ bonds postulated in SO_2—a very significant difference as indicated in Table 6-1. The total atomization energy of SO_3 based on this model is 341.6 compared to 340.0 kpm for the experimental value.

Sulfuric acid can be adequately described as consisting of two OH bonds,

TABLE 6-1

SUMMARY OF HYDRATION OF SULFUR OXIDES

Sulfur Dioxide

$H_2O(g) + SO_2(g) \rightarrow H_2SO_3(g)$

2 O—H	223.6 $\rightarrow$ 2 O—H	227.2	– 3.6
S″=O″	146.4 $\rightarrow$ 2 S′—O′	139.6	+ 6.8
S′—O‴	107.9 $\rightarrow$ S′—O‴	108.7	– 0.8

Net change = heat of hydration   + 2.4
Estimated free energy change   + 7 kpm

Sulfur Trioxide

$H_2O(g) + SO_3(g) \rightarrow H_2SO_4(g)$

2 O—H	223.6 $\rightarrow$ 2 O—H	225.6	– 2.0
S″=O″	146.4 $\rightarrow$ 2 S′—O′	139.6	+ 6.8
2 S″—O″	195.2 $\rightarrow$ 2 S′—O‴	217.4	– 22.2

Net change = heat of hydration   – 17.4
Estimated free energy change   – 12 kpm

two S—OH bonds, and two S′—O‴ bonds. The energies of these add up to 582.6 kpm. The experimental atomization energy of $H_2SO_4(g)$ is reported to be 586.2 kpm. This molecule can consist of two ordinary single bonds to sulfur and two coordinate covalent bonds using the two lone pairs on the sulfur. There is no need to invoke the use of outer d orbitals on the sulfur except in their important role of reducing the bond weakening in the oxygen.

The formation of "orthosulfuric acid," $S(OH)_6$, however, would require

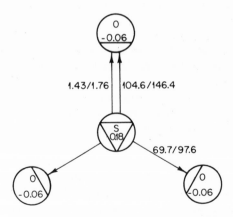

FIG. 6-9. Schematic representation of sulfur trioxide (all bonds equal). Atomization energy: 341.6 kpm, calc.; 340.0 kpm, exp.

promotion of lone pair electrons to outer d orbitals. The latter are not sufficiently stable until the partial charge on the atom is reasonably high. In $S(OH)_6$ the charge on sulfur would be only 0.04. This is presumably the explanation of the fact that $S(OH)_6$ is not known.

Problems of Selenium–Oxygen Chemistry

Selenium and sulfur are so nearly alike in electronegativity (4.12 and 4.21) that the differences in their oxygen chemistry, which are quite striking, are

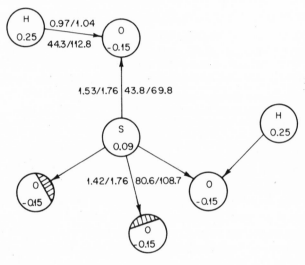

FIG. 6-10. Schematic representation of sulfuric acid. Atomization energy: 582.6 kpm, calc.; 586.2 kpm, exp.

puzzling. At least some light can be shed on these differences if it is assumed that selenium atoms, being larger than those of sulfur, are even less able to form stable double bonds to oxygen. One of the most obvious puzzles is why selenium dioxide is a polymeric solid instead of being gaseous like SO_2. The structure of the solid, however, is just what would be expected if the simple molecule, like SO_2, had one double bond (on the average) and one coordinate covalent bond $Se\text{—}O'''$, but polymerized through the double bond. This would lead to the observed long chains of alternate selenium and oxygen atoms with one oxygen bonded to each selenium only by a $Se\text{—}O'''$ bond. Data for the standard heat of formation of $SeO_2(g)$ are not available, but by assuming the above model of the polymer, one can calculate the heat of formation of the gaseous polymer to be -30.6 kpm. Assuming a model similar to the SO_2 molecule, one can also calculate the standard heat of formation of $SeO_2(g)$ unpolymerized. This is -46.5 kpm. One may conclude tentatively that the enthalpy of polymerization in the vapor state is $+15.9$ kpm, showing the monomer to be favored. The reported standard heat of formation of the solid polymer is -53.9 kpm. This allows only -7.4 kpm for the polymerization from vapor monomer to solid polymer. However, the calculated heat of formation of the gaseous monomer may be too high because it was based on postulating the equivalent of one double bond, which may not be possible. A lower heat of formation of the monomer would of course favor the polymer even more.

In contrast, calculations show that polymerization of sulfur dioxide would have an enthalpy of $+9.2$. This would correspond to an even higher positive free energy change since the entropy decreases with polymerization.

The observed ease of depolymerization of SeO_2, which sublimes readily, is consistent with the low enthalpy of depolymerization. But the state of the bonding in $SeO_2(g)$ must be such, if a stable double bond cannot be present, that the molecule can be kinetically very active as an oxidizing agent, having very low energy of activation. This constitutes another very significant difference between SO_2 and SeO_2. In fact sulfur dioxide readily reduces SeO_2 to elemental selenium.

Another striking difference between SeO_2 and SO_2 is the fact that although sulfurous acid is very unstable, with respect to the anhydride and water, SeO_2 actually absorbs water from the air to form H_2SeO_3. An approximate calculation shows the standard heat of formation of $H_2SeO_3(g)$ to be about -99.7 kpm. The sum of the heats of formation of $SeO_2(g)$ (calculated) and $H_2O(g)$ is -93.3, indicating that even in the gaseous state, the hydration process would be exothermic by about -6.4 kpm. From the experimental heats of formation of solid SeO_2, gaseous water, and solid H_2SeO_3, the hydration process is found to be exothermic by -13.7 kpm.

Although as expected, selenium trioxide has about the same degree of acidity as sulfur trioxide, it is much less stable, losing oxygen above $180°C$. Correspondingly, it is a powerful oxidizing agent, being readily reduced to elemental selenium. A thorough knowledge of the nature of sulfur–oxygen and selenium–oxygen bonds should reveal the cause of this difference. The loss of oxygen from SeO_3 is easily explained, since from the heats of formation of solid SeO_2 and SeO_3 one can calculate that the loss of oxygen from SeO_3 to leave SeO_2 is exothermic by -14.0 kpm. Approximate calculations show that in an SeO_3 polymer, assumed to have the structure of an SO_3 polymer, the terminal oxygen atoms must be attached by Se—O″ bonds instead of the Se—O‴ bonds existing in SeO_2. This in part would account for the exothermicity of the oxygen loss.

Tellurium

Thermochemical data on tellurium are sparse. It would be interesting to develop a quantitative explanation of the fact that telluric acid is the ortho acid, $Te(OH)_6$ rather than H_2TeO_4. Presumably this is related to the inability of tellurium to reduce the "lone pair weakening" of oxygen, thus favoring the maximum possible number of single O' bonds. But since iodine does form HIO_3, which is probably polymeric, further study is obviously needed.

Chlorine and Oxygen

Not many data are available for oxyhalogen compounds either. The monoxide ClO is reported to have a dissociation energy of 64.5 kpm, substantially higher than the Cl—O' energy of 54.0. The average of a Cl—O'' and a Cl—O' energy would be 63.6 kpm, but there is no reason to expect this kind of a bond. Possibly the experimental value is in error. For Cl_2O, the calculated and experimental atomization energies are in exact agreement at 98.6 kpm. Here the ionic weighting coefficient is 0.03, leading to a covalent contribution of 43.4 and an ionic contribution of 5.9 kpm.

Data for gaseous perchloric acid, $HClO_4$, are not available. The standard heat of formation of the liquid is reported as -9.70 kpm. This provides a guide to the kind of bond in the molecule, for the heat of formation of the gas must be somewhat less than -9.7. It would not be less if the ClO bonds were stronger than Cl—O' bonds. On the basis of that assumption, a standard heat of formation for the gas is calculated to be $+9.9$ kpm, which certainly seems at least approximately consistent with the value reported for the liquid. One may conclude that chlorine in this compound is not capable of reducing the lone pair weakening on oxygen, which forms coordinate covalent bonds to the chlorine lone pairs. The partial charge on chlorine, -0.02, would not warrant expectation of availability of its outer d orbitals.

Chapter 7

APPLICATIONS TO INORGANIC MOLECULES

II. HALIDES

Alkali Halide Gas Molecules

Bond energy calculations for the alkali halide gas molecules are summarized in Table 7-1. It may be observed that the poorest agreement between calculated and experimental values occurs in the fluorides and that the

TABLE 7-1

BOND ENERGIES OF ALKALI HALIDE GAS MOLECULES

Compound	R_c (Å)	R_o (Å)	t_i	E_c	E_i	Atomization energy (kpm) Calc.	Exp.
LiF	2.05	1.55	0.74	11.2	158.5	169.7	136.8
LiCl	2.33	2.03	0.65	16.3	106.3	122.6	114.3
LiBr	2.48	2.17	0.61	16.1	93.3	109.4	101.9
LiI	2.67	2.39	0.53	16.8	73.6	90.4	85.7
NaF	2.25	1.85	0.75	7.9	134.6	142.5	114.8
NaCl	2.53	2.36	0.67	11.5	94.3	105.8	99.0
NaBr	2.68	2.50	0.62	11.7	82.3	94.0	88.8
NaI	2.87	2.71	0.54	12.4	66.2	78.6	72.2
KF	2.67	2.14	0.85	4.2	131.9	136.1	118.1
KCl	2.95	2.67	0.76	7.4	94.5	101.9	101.6
KBr	3.10	2.82	0.71	7.9	83.6	91.5	91.0
KI	3.29	3.05	0.63	8.7	68.6	77.3	77.8
RbF	2.87	2.24	0.86	4.0	127.5	131.5	115.7
RbCl	3.15	2.79	0.78	6.7	92.8	99.5	100.5
RbBr	3.30	2.95	0.73	7.2	82.2	89.4	90.4
RbI	3.49	3.18	0.65	8.1	67.9	76.0	76.8
CsF	3.06	2.35	0.90	2.6	127.1	129.7	116.2
CsCl	3.34	2.91	0.81	5.5	92.4	97.9	101.
CsBr	3.49	3.07	0.77	5.8	83.3	89.1	91.
CsI	3.68	3.32	0.69	6.8	69.0	75.8	75.

next poorest agreement is in the other halides of lithium and sodium. In all of them the calculated value is significantly higher than the experimental value. In total, eleven molecules show excessive deviation between calculated and experimental dissociation energy.

From potassium on, all the chlorides, bromides, and iodides have experimental bond energies in good agreement with the calculated values. Here are nine molecules where results are quite satisfactory. Why should the method be accurately applicable to nine out of twenty such similar molecules, and not to the others? Similar calculations applied to all twenty of these same compounds in the solid state give good agreement throughout. Although gas phase data for other highly polar halide molecules are not abundant, insofar as they exist they are in good agreement with the calculated values. The cause of the discrepancies among the alkali halide gas molecules is not known. Various difficulties are encountered in the experimental measurement of bond energies of these compounds in the gas phase. Not least of these is the problem resulting from some degree of association in the vapor. Fluorides, and the compounds of the smaller atoms of lithium and sodium, might be expected to give greatest difficulty of this type, because the electrostatic interaction between molecules would be greater for the smaller atoms where the distances are shorter.

It is of course frequently claimed that these molecules are all ion pairs in the gas phase and completely ionic. Deviations from the purely electrostatic bond energies are then ascribed to polarization effects. Indeed, the bond energies of these compounds have been successfully calculated[1] purely from such considerations and based on the totally ionic model. But if the polar covalent model is accepted, then some other cause of the discrepancies probably must be found. Conclusive arguments either way are apparently not yet available.

Hybridization in the Gaseous Halides

Since the outermost shell of each atom consists of one s and three p orbitals, the s orbital being more stable than the p, it has usually seemed important to take these differences into account in discussion of bond energy. Indeed, the literature of theoretical chemistry contains very numerous references to promotional energies and hybridization and the alleged effects of differing degrees of "s character" in the hybridization. It is therefore of interest that such effects have presented little problem in the bond energy work described herein. Table 7-2 summarizes all available data on gaseous mono-, di-, tri-, and tetrahalides of the same elements. For elements of group M2, the ground state electronic structure includes an outermost shell with the two valence

[1] E. S. Rittner, *J. Chem. Phys.* **19**, 1030 (1951).

TABLE 7-2

EFFECT OF NUMBER OF BONDS ON BOND ENERGY

Compound	Atomization energy (kpm)		Compound	Atomization energy (kpm)		Compound	Atomization energy (kpm)	
	Calc.	Exp.		Calc.	Exp.		Calc.	Exp.
BeF'''	150.1	146.9	BF'''	150	181	AlCl'''	98.7	84.8
BeF$_2$'''	304.8	303.6	BF$_2$'''	302.4	302.3	AlCl'''$_3$	307.5	305.0
			BF$_3$'''	462.9	463.0			
MgF''	130.6	132.0				SiF''''	132.3	132.
MgF$_2$'	262.2	258.9	BCl'''	102.6	120.4	SiF$_2$''''	289.0	289.
			BCl$_2$'''	211.0	213.4	SiF$_3$''''	423.0	433.1
MgCl'	97.4	71±16	BCl$_3$'''	316.2	318.3	SiF$_4$''''	568.4	570.5
MgCl$_2$'	203.2	204.3						
			BBr'''	88.3	99.6	SiCl'''	95.0	92.6
MgBr'	84.6	82±7	BBr$_2$'''	179.0	176.8	SiCl$_2$'''	191.8	206.7
MgBr$_2$'	177.2	173.4	BBr$_3$'''	268.5	264.3	SiCl$_4$'''	380.0	382.3
CaF''	124.3	127±3	BI'	63.8	81.7	NF'	68.9	73.3
CaF$_2$''	265.4	268.3	BI$_2$'	127.6	131.0	NF$_2$'	137.8	140.7
			BI$_3$'	197.7	194.1	NF$_3$'	206.7	200–2
SrF''	124.4	127±3						
SrF$_2$''	268.8	262.8	AlF'''	132.2	158.2	PF'''	114.7	112.7
			AlF$_2$'''	275.8	272.8	PF$_2$'''	237.8	222.6
BaF''	125.6	136±3	AlF$_3$'''	420.0	421.2	PF$_3$'''	363.3	356.1
BaF$_2$''	273.4	273±4						
						PBr'	61.9	63.5
						PBr$_3$'	190.8	193.3

electrons in the s orbital. Such atoms can form not even one bond unless one of these electrons is promoted to a p orbital. Any promotional energy is thus the same for the monohalides as the dihalides. Evidently it is adequately taken care of in the evaluation of the electronegativity and homonuclear single covalent bond energy parameter for Table 7-2 discloses no significant difficulties in calculating the bond energies of both mono- and dihalide molecules of these elements.

For boron and perhaps aluminum the situation is somewhat different. Here the ground state is represented by an outer shell of structure s^2p. No promotion is necessary for univalence. Here we observe that the experimental bond energy is significantly higher than the calculated value for the monohalides, except for AlCl which may be in error. Two possible explanations suggest themselves. One is that retaining the lone pair would reduce the electronegativity of the atom, below its "normal" value, thus increasing the bond polarity and raising the energy from this cause. The other is that in the M3 elements there are two vacant orbitals in addition to the lone pair and the bonding orbital. The presence of these might encourage a small degree of multiplicity which would increase the bonding energy.

The data on M4 compounds allow a choice of these explanations by eliminating the first. For if the monohalides of boron have abnormally high bond energy because of a lone-pair effect, then both the monohalides and dihalides of silicon should exhibit the same effect. The tabulated data give no evidence of such an effect nor, indeed, of any significant difference among mono-, di-, tri-, and tetrahalides. The compounds of nitrogen and phosphorus likewise reveal no significant difference associated with the number of bonds.

The data on M4 compounds suggest that hybridization occurs whether or not a lone pair exists. Otherwise one might expect to see a difference between the silicon–halogen bonds depending on whether the arrangement of the silicon atom is s^2pp or sp^3.

There seems to be ample evidence that utilization of lone pairs in forming coordinate covalent bonds has no significant effect on the parameters used for calculating energies of other bonds to the same atom. The phosphorus to halogen bonds, for example, are evidently the same in the trihalides as they are in the oxyhalides. Phosphorus trifluoride consists of three P'—F''' bonds, with a total atomization energy of 363.3 kpm, compared to the experimental energy of 357.9 kpm. In phosphorus oxyfluoride, POF_3, the P'—F''' bonds are the same and a P'—O''' bond is added, giving a total atomization energy of 489.7 kpm, compared to the experimental value of 481.0.

It is relevant to examine here the effect of promotion to an outer d orbital. This appears to be a phenomenon much less common than many chemists, including the present author, once believed. One interesting example is that of $SF_4(g)$. Presumably the formation of this molecule would require the promotion of one of the sulfur lone pairs out to an outer d orbital, permitting the formation of two new covalent bonds. Yet the total energy of four S'—F' bonds in this molecule is 327.2 kpm, compared to the experimental atomization energy of 327.4 kpm. The homonuclear single covalent bond energy parameter of sulfur used in this calculation is the 55.0 derived from the experimental heat of formation of hydrogen sulfide. There is no evidence here that promotion of one of the lone-pair electrons under these conditions requires an appreciable amount of energy, unless this amount is then balanced almost exactly by an increase in all the bond energies, associated with a new degree of hybridization.

We may assume that in SF_4, the remaining lone pair is s^2. This implies that one of a p^2 pair was promoted to create the two new bonding orbitals. There is evidence that promotion of one of the s^2 electrons may require a significant net amount of energy, because the calculated atomization energy of SF_6, 490.8 kpm, is 19 kpm higher than the experimental value of 471.8 kpm. However, this is not supported by the experimental atomization energy of SeF_6, which is 430 kpm compared to the calculated value of 422.4. Similarly, the experimental atomization energy of TeF_6, 474 kpm, is appreciably higher than the calculated value of 459.6 kpm.

As mentioned above, phosphorus trifluoride can be described as having three P'—F''' bonds. The pentafluoride appears to be formed by creating two new P'—F' bonds. The atomization energy calculated on this basis is 554.4 kpm. The experimental value is 555.1 kpm. No significant promotional energy appears to be involved here. But the situation is clearly different with the chlorides. The phosphorus trichloride molecule seems to be adequately described as consisting of three P'—Cl' bonds, giving a total calculated atomization energy of 225.3 kpm, to be compared with the experimental value of 231.1 kpm. However, the pentachloride molecule has an experimental atomization energy much too low for even five P'—Cl' bonds. If the energy of three P'—Cl'' bonds, not as in PCl_3, is calculated, and added to the energy of only one P'—Cl' bond, however, the total is 311.0 kpm, close to the experimental value of 310.3 kpm. The lengths of the P'—Cl'' bonds are 2.04 Å, compared to the covalent radius sum of 2.09 Å and the two new bonds of 2.19 Å. These data suggest that there is no promotion of the lone pair in PCl_3 to form PCl_5. Rather, two half-bonds form. This kind of bonding is quite analogous to the molecular orbital description proposed for such compounds as XeF_2. It may be more common than generally supposed. In the following section it will be examined further.

Interhalogen Compounds

In the past, polyhalogens such as the tri- and pentafluorides have conveniently been explained as resulting from promotion of lone pair electrons from the less electronegative halogen to outer d orbitals. This provided a ready explanation of the observation that only one, three, five, or seven fluorine atoms can become attached to the other halogen atom. No halogen difluoride or tetrafluoride is known. This is consistent with the fact that the valence starts with one, from the normal electronic configuration of the halogens, and must increase by two's because promotion of one electron creates two new bonding orbitals simultaneously. However, a study of the bond energies involved in such compounds shows that in no instance is the atomization energy high enough to account for the full number of single covalent bonds. Limited promotion to d orbitals apparently does occur, but not enough to account for all the bonds. It appears useful to postulate "half-bonds," which occur only in pairs and share a normal single covalent bond energy between them.

TABLE 7-3

BONDING IN INTERHALOGEN COMPOUNDS

Compound	Bond	R_0 (Å)	t_i	E_c	E_i	$E_{calc.}$	E
ClF	Cl'—F'	1.63	0.08	45.0	16.3	61.3	59.9
ClF$_3$	Cl'—F'	1.60	0.09	45.3	18.7	64.0	
	Cl'—F'	1.70	0.09	42.7	17.6	60.3 (2 half-bonds)	
Total						124.3	124.7
ClF$_5$	Cl'—F'	1.62	0.09	44.8	18.4	63.2	
	Cl'—F'	1.72	0.09	42.2	17.4	119.2 (4 half-bonds)	
Total						182.4	181.4
BrF	Br'—F'	1.76	0.13	38.2	24.5	62.7	59.6
BrF$_3$	Br"—F"	1.72	0.13	56.2	25.1	81.3	
	Br'—F'	1.81	0.13	37.2	23.8	61.0 (2 half-bonds)	
Total						142.3	144.5
BrF$_5$	Br'''—F'''	1.68	0.13	73.2	25.7	98.9	
	Br'—F'	1.78	0.13	37.8	24.2	124.0 (4 half-bonds)	
Total						222.9	223.7
BrCl	Br'—Cl'	2.14	0.04	49.4	6.2	55.6	51.3
IF	I'—F'	1.91	0.21	31.2	36.5	67.7	67.0
IF$_5$	I'''—F'''	1.75	0.22	60.4	41.7	102.1	
	I'—F'	1.86	0.22	31.7	39.2	212.7 (2 bonds + 2 half-bonds)	
Total						314.8	320.0
IF$_7$	I'''—F'''	1.75	0.22	60.4	41.7	102.1	
	I'—F'	1.86	0.22	31.6	39.3	283.8 (2 bonds + 4 half-bonds)	
Total						385.9	386.9
ICl	I'—Cl'	2.30	0.13	40.3	18.8	59.1	50.3
IBr	I'—Br'	2.45	0.08	37.8	10.8	48.6	42.4

Table 7-3 summarizes in some detail the bond energy calculations for the interhalogen compounds for which experimental data are available. It is interesting that all the polyfluorides appear to consist of one relatively strong bond and other weaker bonds. This is consistent with the observed structural data, which include one shorter bond. The strong bond is a short Cl'—F' bond in ClF$_3$ and ClF$_5$, a Br"—F" bond in BrF$_3$ and a Br'''—F''' bond in BrF_5, and an I'''—F''' bond in IF_5 and IF_7. Then ClF_3 appears to have two other longer bonds with a total energy of one Cl'—F' bond. BrF_3 similarly has two other longer bonds with the total energy of one Br'—F' bond. In both ClF_5 and BrF_5 the other four bonds also appear to be half-bonds, having the total energy of two normal single X'—F' bonds. In IF_5, however, two of the other bonds average full single bonds, and the other two are half-bonds. In IF_7, the bonds resemble those of IF_5 except that two new half-bonds are added. Unfortunately, the molecular structures of these compounds do not appear to be accurately enough known to identify, with certainty, minor bond length differences, such as might occur if the long half-bonds and full bonds are not completely averaged.

It is only fair to point out that there is as yet no theoretical basis for predicting the applicability of double and triple prime bond energy parameters as postulated above. One can only note some consistency in the occurrence of higher prime single bond energies with higher positive charge on the central atom. Nevertheless the agreement between experimental atomization energies and those calculated for the structures postulated in Table 7-3 is quite impressive.

As previously mentioned, the xenon fluorides have been explained as involving a type of half-bonding, as an alternative to the promotion of outer lone-pair electrons to outer d orbitals. The half-bonds are proposed as forming in pairs at opposite sides of the central atom by interaction of three atomic orbitals and four electrons. The orbitals are provided one by the central atom and one by each fluorine. They form three center molecular orbitals, one bonding, one nonbonding, and one antibonding. The four electrons come one from each fluorine atom and two from a lone pair of the central atom. Two fill the bonding orbital and the other two occupy the nonbonding orbital. Since two electrons in the bonding molecular orbital are holding two fluorine atoms to the central atom, the bonds are equivalent to half-bonds. It is very interesting that the quantitative calculations show such half-bonds to be so close to half the energy of the normal two electron single bond.

To the limited extent that a tentative knowledge of the nature of the bonding in these compounds provides a quantitative explanation of their stability, one can now understand why bromine and iodine, but not chlorine, favor the higher fluorides over the monofluorides. The experimental standard heats of formation (which are not significantly different from the calculated values) are as follows: ClF, -11.9; ClF_3, -38.9; ClF_5, -57.8; BrF, -14.0; BrF_3, -61.1; BrF_5, -102.5; IF, -22.6; IF_5, -200.0; IF_7, -229.1 kpm. From these values together with the standard heats of formation of $Br_2(g)$ $+7.4$ and $I_2(g)$ $+14.9$, one can calculate enthalpies of the following gas state reactions:

$$3 \text{ ClF} \rightarrow \text{ClF}_3 + \text{Cl}_2 \qquad \Delta H^0 = -3 \text{ kpm}$$
$$5 \text{ ClF} \rightarrow \text{ClF}_5 + 2 \text{ Cl}_2 \qquad \Delta H^0 = +2 \text{ kpm}$$
$$3 \text{ BrF} \rightarrow \text{BrF}_3 + \text{Br}_2 \qquad \Delta H^0 = -11.7 \text{ kpm}$$
$$5 \text{ BrF} \rightarrow \text{BrF}_5 + 2 \text{ Br}_2 \qquad \Delta H^0 = -17.7 \text{ kpm}$$
$$5 \text{ IF} \rightarrow \text{IF}_5 + 2 \text{ I}_2 \qquad \Delta H^0 = -57.2 \text{ kpm}$$
$$7 \text{ IF} \rightarrow \text{IF}_7 + 3 \text{ I}_2 \qquad \Delta H^0 = -26.2 \text{ kpm}$$

All these reactions would be accompanied by entropy decrease, which would probably change the sign of the free energy change of the first reaction and increase the positive free energy of the second. It would have relatively little effect on the marked tendency of the lower bromine and iodine fluorides to disproportionate to the higher ones and the free bromine or iodine.

Miscellaneous Problems

CARBONYL HALIDES

The calculated atomization energy of phosgene, $COCl_2$, assuming an ordinary $C=O''$ double bond and two single $C-Cl'$ bonds, is 366.9 kpm, whereas the experimental value is only 341.4 kpm. If the CO bond were an average of a $C=O''$ bond and a $C-O'''$ bond, then the total atomization energy would be 343.4, in agreement with the experimental value. This kind of a bond is not unknown, being observed in the sulfur oxides and in gaseous SiO_2. But since it has not previously been recognized in a compound of carbon, a further check is needed. Such a check is offered by carbonyl fluoride.

There is evidence that the first two fluorines attached to carbon may be considered $C-F''$ bonds despite the apparent absence of stable empty orbitals on the carbon atom. This evidence is tabulated in Table 7-4. On the

TABLE 7-4

EVIDENCE FOR $C-F''$ BONDS IN CARBON COMPOUNDS[a]

Compound	Atomization energy (kpm)	
	Calc.	Exp.
CF_2	250.2	250.
CF_4	458.8	467.9
CF_2Cl_2	379.8	382.3
COF_2	418.2	420.4

[a] A comparison of calculated with experimental atomization energies assuming two $C-F''$ bonds per molecule.

basis of this evidence, it seems reasonable to assume that the two $C-F$ bonds in COF_2 are also of the $C-F''$ type. However, the atomization energy calculated for COF_2 in the table is also based on the assumption that the CO bond, as in $COCl_2$, is an average of $C\!=\!O''$ and $C\!-\!O'''$.

As will be seen in a later chapter, this may also be the nature of the carbonyl bond in formic acid. In general, carbon which forms two single bonds and one double bond seems to tend to average the double bond with a coordinate covalent bond, carbon donor, when the partial charge on carbon is relatively high, i.e., 0.10 or higher. This of course is only a very tentative generalization, of which only a few possible examples exist.

If the phenomenon of double prime and triple prime single bonding usually depends, as implied in Chapter 3, on the availability of vacant orbitals in the other atom, we may well ask the question: Why does it seem to occur when fluorine is attached to carbon? Perhaps carbon in a highly oxidized condition does have the ability to use orbitals in the next (third) principal quantum level.[2] However this does not apply when all the bonds to the carbon atom are single bonds, as in hexafluoroethane, wherein all CF bonds appear to be $C\!-\!F'$. The calculated and experimental atomization energies of C_2F_6 are 692.1 and 696 kpm, respectively.

CARBON TETRACHLORIDE

Although the carbon–chlorine bonds are strong enough to hold this molecule together despite the crowding, crowding does exist and must lower the total bonding energy. The internuclear distance between chlorine atoms of the same molecule is less than the usual van der Waals distance, which means that CCl_4 cannot form without distortion of the electronic clouds of the chlorine atoms. Support of this statement is to be found in the bond length, 1.77 Å, which for an appreciably polar bond ($t_i = 0.15$) should be appreciably shorter than the covalent radius sum, 1.76 Å. The experimental atomization energy of carbon tetrachloride is only 312.3 kpm. The calculated value, neglecting the chlorine–chlorine repulsions through steric crowding, is 347.6 kpm. The total bonding energy is thus weakened by about 35 kcal. As will be seen later, it is therefore erroneous to consider one-fourth of the atomization energy of CCl_4 as fairly representing a typical C—Cl bond energy.

Two halogen atoms can be tolerated reasonably well on the same carbon atom, but not three if they are chlorine or larger. For chloroform, $CHCl_3$, for example, the calculated atomization energy is 346.4 kpm compared to the experimental value of 334, despite the possibility that the energy may be enhanced somewhat by internal protonic bridging. The calculated atomization energy of $CFCl_3$ is 379.6 compared to the experimental value of 345.5 kpm. In contrast, the calculated value for CF_2Cl_2, 379.8, agrees reasonably well with the experimental value of 382.3 kpm.

[2] E. A. C. Lucken, *J. Chem. Soc.* p. 2954 (1959); J. F. A. Williams, *Trans. Faraday Soc.* **57**, 2089 (1961).

In the light of the above evidence, it should not be surprising that higher chlorocarbons are unstable and that CBr_4 and CI_4 are little known and tend to give tetrahaloethenes on decomposition. Nor should it be surprising that although hydrolysis of CCl_4 is thermodynamically favored at ordinary temperatures, it is mechanistically unfeasible.

NITROGEN HALIDES

Judging from literature discrepancies the standard heat of formation of $NF_3(g)$ is not accurately known. However, the consensus seems to favor a value about -31 and the calculated value is -37.0 kpm. In contrast, the standard heats of formation of NCl_3 and NBr_3 are calculated to be $+35.3$ and $+44.3$ kpm. One can thus account for the relative stability of NF_3 and the relative instability of the higher halides. The violence of the explosive decomposition of the higher halides is of course a matter of kinetics, plus the gaseous character of the products.

SULFUR DIFLUORIDE

One of the puzzling aspects of sulfur chemistry is the fact that SF_2 is scarcely known whereas SF_4 and SF_6 are commercial substances. Superficially, at least, one would expect SF_2 to be a very common substance, since the two half-filled orbitals of the sulfur atom are adequately utilized by the two fluorine atoms. This compound has only recently been positively identified by Johnson and Powell using microwave spectroscopy.[3] It is indeed elusive and has only a brief existence. Calculation shows that the atomization energy of SF_2 is 157.6 kpm, which corresponds to a standard heat of formation of -43.2 kpm. The compound is therefore thermodynamically stable in the usual sense of the word but it must be very easy for it to disproportionate to substances having even greater stability. In fact, one can calculate for the change of 2 moles of SF_2 to solid sulfur and $SF_4(g)$ that the enthalpy change is -98.8 kpm. Here then is another example of a stable compound which is elusive and little known not because it is inherently weakly bonded, but because it is capable easily of change to more stable substances. The energy of an S—F bond in SF_2 is 78.8 and in SF_4, 81.8 kpm, so the advantage of this change is relatively small. The chief contribution to the energy is the liberation of free sulfur in the reaction.

SULFUR IODIDE

Another idiosyncrasy of sulfur is its apparent inability to form stable bonds to iodine. Although not very stable, the chlorides and bromides of sulfur are well known.

A simple bond energy calculation for sulfur diiodide is revealing. The

[3] D. R. Johnson and F. X. Powell, *Science* **164**, 950 (1969).

sulfur–iodine nonpolar energy is 44.6 kpm. From electronegativity equalization, the partial charges are found to be low, -0.04 for sulfur and $+0.02$ for iodine, leading to $t_i = 0.03$. The bond length would not be very different from the covalent radius sum of 2.37 Å. The covalent energy contribution is then

$$t_c E_c = \frac{0.97 \times 44.6 \times 2.37}{2.37} = 43.3 \text{ kpm}$$

The ionic energy contribution is

$$t_i E_i = \frac{0.03 \times 332}{2.37} = 4.2 \text{ kpm}$$

The sum is 47.5 kpm, so the atomization energy of the gaseous SI_2 is 95.0 kpm.

The energy to atomize from their standard states 1 mole of sulfur atoms and 2 moles of iodine atoms is $66.6 + 51.0 = 117.6$ kpm. Clearly the atoms are more stably held in the free elements than in the compound. The calculated standard heat of formation of sulfur diiodide gas is $117.6 - 95.0 = +22.6$ kpm. The nonexistence of sulfur iodides no longer seems mysterious. Evidently any known reaction that might produce a sulfur–iodine bond can more easily liberate iodine and sulfur or undergo some other change that is energetically more favorable.

CYANOGEN HALIDES

A study of the bond energies in cyanogen halides reveals that the experimental atomization energies are too low to provide for a normal triple bond between carbon and nitrogen. They suggest, instead, that the bond can be represented as C≡N‴, which has an energy about 45 kcal lower than would be expected of C≡N‴. Whether this indicates that the cyanide bond is really only a double bond to a triple prime nitrogen, or whether for some reason the triple bond multiplicity factor is only 1.50 for the cyanide bond is not known. However, the following data are obtained, calculations being based on the assumption of a normal C—X′ single bond to halogen and a C≡N‴ bond. The calculated and experimental atomization energies are FCN, 306.0, 305; ClCN, 288.4, 280.4; BrCN, 270.9, 266.5; ICN, 248.7, 255.9 kpm. For HCN they are 291.2 and 304.1 kpm.

This unusual observation is consistent with the apparent condition of the bonds in organic nitriles, at least as evidenced by acetonitrile, CH_3CN. Assumption of a C≡N‴ bond here, instead of a triple bond, leads to a calculated atomization energy of 588.7 kpm, in good agreement with the experimental value of 590.9 kpm. Assumption of an acetylenic triple bond would have given a calculated atomization energy 35 kpm too high.

Xenon Tetrafluoride

Extrapolation of nonpolar covalent radii leads to a value of 1.31 Å for xenon, from which an electronegativity of 4.06 has been estimated.[4] From this value one can calculate an electronegativity of XeF_4 that corresponds to partial charges of 0.31 on xenon and -0.08 on fluorine, or, $t_i = 0.20$.

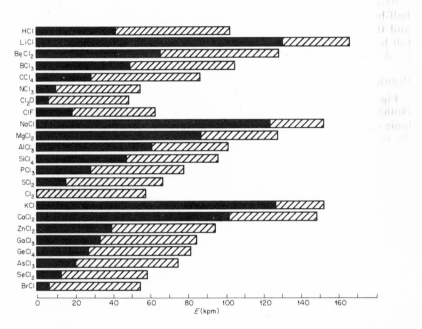

FIG. 7-1. Periodicity of chloride bond energies. Black areas represent ionic contribution.

A homonuclear single covalent bond parameter for xenon can be estimated by extrapolation: Sb, 32; Te, 34; I, 36; Xe (38). The geometric mean with fluorine is 37.9 kpm. The observed bond length is 1.95 Å compared to the nonpolar covalent radius sum of 2.02 Å. Considering the bonds as half-bonds having half the energy of a normal covalent bond, one can then calculate the covalent contribution as

$$t_c E_c = \frac{0.50 \times 0.80 \times 37.9 \times 2.02}{1.95} = 15.7 \text{ kpm}$$

The ionic contribution is

$$t_i E_i = \frac{0.50 \times 0.20 \times 332}{1.95} = 17.0 \text{ kpm}$$

[4] R. T. Sanderson, *J. Inorg. Nucl. Chem.* **27,** 989 (1965).

The sum, 32.7 kpm, represents the average bond energy in XeF_4. It also corresponds to an atomization energy of 130.8 kpm and a standard heat of formation calculated as -55.2 kpm.

A number of bond energy estimates based on experimental data have agreed with an average bond energy of about 32 kpm. The National Bureau of Standards Technical Note 270–1 (1965) lists only the heat of formation of solid XeF_4 as -62.5 kpm. From the reported heat of sublimation of 15.3 kpm this corresponds to a standard heat of formation of the gas of -47.2 kpm and an average bond energy of 30.7 kpm. It appears that the half-bond postulate effective in the calculation of bond energies in PCl_5 and the halogen fluorides is consistently applicable to the xenon fluorides (all have nearly the same energy as would also be calculated), and, as will be shown later, to the bifluoride ion.

PERIODICITY OF CHLORIDE BOND ENERGIES

Figure 7-1 shows the periodicity of bond energies in some binary chlorides. Notice that although the covalent contributions vary somewhat, it is the ionic contributions that are dominant in determining the general nature of the periodic trends observed.

Chapter 8

APPLICATIONS TO INORGANIC MOLECULES

III. HYDROGEN COMPOUNDS AND MISCELLANEOUS COMPOUNDS

Bonding in Diatomic Gas Molecules of Polyvalent Atoms

Use of the bond energy calculation method of identifying bond type is of course least ambiguous when applied to diatomic gas molecules. Although some of such molecules have already been considered in another context, it is interesting to examine them all together. Assuming the experimental dissociation energies to be essentially correct, which may not always be a fair assumption, one can then discover that the bonds in these molecules are not always what one would expect.

Table 8-1 summarizes the data for diatomic gas molecules formed between polyvalent atoms. Although most of these molecules have bonds that would seem easy to predict, there are some interesting deviations. One of the most conspicuous is the BN gas molecule, which one would predict should have a triple bond. The experimental dissociation energy is not nearly large enough. At best it represents, although not very well, a single bond B—N‴. The JANAF reports* give a standard heat of formation of $+152$ kpm. The calculated energy of a BN triple bond is 182.2 kpm. This corresponds to a standard heat of formation of $+65.3$ kpm. The experimental bond length is 1.28 Å compared to the covalent radius sum of 1.56 Å, and is not unreasonable for a triple bond. The fact that the BO molecule does appear to contain a normal double bond suggests that something is amiss in the experimental BN data.

Although too limited to be conclusive, the data appear to indicate a reduced ability to form multiple bonds associated with disparity of size. This does not appear to apply when the less electronegative atom is the smaller (CS), but it does when the more electronegative atom is the smaller. The expected normal triple bonds are observed in N_2, CS, SiS, P_2, and CO,

* See Ref. 1, p. 41.

105

TABLE 8-1

APPARENT BOND TYPES IN DIATOMIC GAS MOLECULES OF POLYVALENT ATOMS

Molecule	Bond	t_i	E_c	E_i	$E_{calc.}$	$E_{exp.}$
C_2	C=C	0	150.6	0	150.6	150.
SiC	Si—C	0.18	64.4	36.2	100.6	103.2
GeC	Ge=C	0.03	100.2	8.1	108.3	110.3
Si_2	Si=Si	0	83.2	0	83.2	75.8
BN	B≡N'''	0.20	138.9	91.9	230.9	92.7
	B—N'	0.20	50.5	51.9	102.4	92.7
SiN	Si—N''	0.21	57.6	44.4	102.0	105.6
N_2	N≡N'''	0	225.8	0	225.8	226.0
PN	P'''=N'''	0.13	117.3	43.5	160.8	155.4
SN	S''=N''	0.04	110.4	13.3	123.7	115.
P_2	P''=P'''	0	125.1	0	125.1	125.1
BeO	Be—O'	0.35	30.5	72.6	103.1	106.9
CaO	Ca—O'	0 57	19.8	104.0	123.8	93±5
BaO	Ba—O'	0.67	13.4	114.7	128.1	130±5
BO	B—O'''	0.27	77.3	74.7	152.0	—
	B=O''	0.27	95.1	112.1	207.2	—
Average					179.6	188.
AlO	Al—O''	0.38	41.5	77.8	119.3	115.8
CO	C≡O'''	0.16	178.4	83.2	261.6	257.3
SiO	Si=O'''	0.29	97.1	95.7	192.8	192.3
GeO	Ge=O'''	0.19	99.9	57.3	157.2	160.6
NO	N''=O''	0.08	117.0	34.6	151.6	151.0
PO	P'—O'''	0.21	70.7	48.1	118.8	120.4
O_2	O''=O''	0	119.1	0	119.1	119.2
SO	S''=O''	0.12	100.4	40.1	140.5	—
	S'—O'''	0.12	76.8	26.7	103.5	—
Average					122.0	124.7
BS	B=S'''	0.15	101.1	46.5	147.6	119.4
CS	C≡S'''	0.04	152.4	15.4	167.8	181.9
SiS	Si≡S'''	0.17	102.3	51.8	154.1	148.6
PS	P'''≡S'''	0.09	98.5	23.4	121.9	123.9
S_2	S''=S''	0	102.5	0	102.5	102.4

but the PO bond seems to be a triple prime single bond, the PN bond a triple prime double bond, and the SiO bond also a triple prime double bond. The reduced ability of sulfur to double bond to oxygen is shown by the bond in SO which appears to average a double and a single bond, as do the bonds in SO_2 and SO_3. Indeed, it is not certain whether a true olefinic double bond exists between sulfur and oxygen in any compound.

Substituted Cyclopolyphosphazenes

Special interest has long been shown in compounds featuring rings that alternate phosphorus and nitrogen atoms, of which perhaps the best known is the trimer of "phosphonitrilic chloride" $(PNCl_2)_3$. This interest has arisen from the presumption that these rings have a pseudoaromatic character presumably associated with pi electrons.[1] Heats of formation of four gaseous derivatives of these cyclopolyphosphazenes have been reported. These

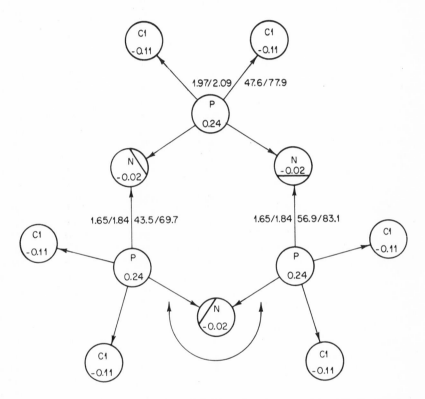

FIG. 8-1. Schematic representation of dichlorophosphazene trimer (all ring bonds are equal). Atomization energy: 925.8 kpm, calc.; 915.1 kpm, exp.

compounds are listed in Table 8-2, together with the usual bond energy calculation data. It is especially interesting to note how well these molecules can be represented as having phosphorus to nitrogen bonds that are the average of two single bonds: P'—N' and P'—N". It is as though each phosphorus atom forms three normal single bonds, two to chlorine atoms and one to a nitrogen atom. It then uses its lone pair to coordinate to another nitrogen atom (see Fig. 8-1). This, being attached to another phosphorus

[1] D. P. Craig and N. L. Paddock, *J. Chem. Soc.* p. 4118 (1962); H. R. Allcock, *Chem. Eng. News* p. 68 (April 22, 1968).

TABLE 8-2

BOND ENERGIES IN SOME SUBSTITUTED CYCLOPOLYPHOSPHAZENES

Compound	Bond	E_{AB}	t_i	E_c	E_i	Bond energy	No. bonds	Total energy (kpm) Calc.	Total energy (kpm) Exp.
$(PNCl_2)_3$	P'—Cl'	54.7	0.18	47.6	30.3	77.9	6	467.4	
	P'—N'	44.8	0.13	43.5	26.2	69.7	3	209.1	
	P'—N"	58.6	0.13	56.9	26.2	83.1	3	249.3	
Total								925.8	915.1
$(PNCl_2)_4$	P'—Cl'	54.7	0.18	47.6	30.3	77.9	8	623.2	
	P'—N'	44.8	0.13	43.5	26.2	69.7	4	278.8	
	P'—N"	58.6	0.13	56.9	26.2	83.1	4	332.4	
Total								1,234.4	1,221.7
$(PN(CH_3)_2)_3$	H—C	91.8	0.03	89.0	9.1	98.1	18	1,765.8	
	C—P'	65.3	0.05	62.7	9.0	71.7	6	430.2	
	P'—N'	44.8	0.12	44.0	24.1	68.1	3	204.3	
	P'—N"	58.6	0.12	57.5	24.1	81.6	3	244.8	
Total								2,645.1	2,637.3
$(PN(C_6H_5)_2)_4$	C—C	83.2	0	122.6	0	122.6	48	5,884.8	
	H—C	91.3	0.03	89.4	9.2	98.6	40	3,944.0	
	C—P'	65.3	0.05	64.1	9.2	73.3	8	586.4	
	P'—N'	44.8	0.12	44.0	24.1	68.1	4	272.4	
	P'—N"	58.6	0.12	57.5	24.1	81.6	4	326.4	
Total								11,014.0	11,000.5

atom by a normal single covalent bond, has two lone pairs left. At least one of these can interact with the phosphorus d orbitals imparting a degree of somewhat delocalized "π" character.

Hydrogen Compounds

The uniqueness of hydrogen resulting from its low atomic number and the absence of electrons below the valence shell causes complications in the calculation of bond energies of hydrogen compounds. A partial charge on a combined hydrogen atom represents a greater, and usually very much greater, percentage change in its electronic cloud than for any other kind of atom. For example, acquisition of a complete electron by a fluorine atom increases its electronic cloud by 11% and, similarly, formation of a chloride ion increases the chlorine cloud by only 6%. In contrast, acquisition of a complete electron by a hydrogen atom produces a 100% increase. Similarly, a 25% change in the electronic cloud of hydrogen would correspond to less than a 3% change in fluorine and only about 1.5% in chlorine.

Compounds of Positive Hydrogen

Early bond energy calculations involving compounds in which hydrogen atoms bear partial positive charge gave results of the right order of magnitude, but consistently somewhat too high. It was reasoned that the nonpolar covalent bond energy contribution of hydrogen must be reduced to the

TABLE 8-3
Atomization Energies of Binary Hydrogen Compounds

Compound	E_{AB}	δ_H	t_i	E_c	E_i	Atomization energy (kpm)	
						Calc.	Exp.
CH_4	92.8	0.01	0.03	90.0	9.1	396.4	397.6
GeH_4	71.7	0	0	71.7	0	286.8	276.7
NH	60.3	0.11	0.11	54.7	35.1	89.8	85.9
NH_2	61.6	0.07	0.11	56.4	35.5	183.8	176.9
NH_3	62.3	0.05	0.11	57.6	35.8	280.2	280.3
N_2H_4	61.7	0.07	0.11	57.1	35.8	411.6	411.6
HN_3	58.3	0.17	0.12	53.3	39.1	316.8	320.8
PH	73.1	− 0.01	0.01	72.4	2.3	74.7	72.7
PH_2	73.1	− 0.01	0.01	72.4	2.3	149.4	154.9
PH_3	73.1	− 0.01	0.02	71.6	4.7	228.9	234.8
P_2H_4	73.1	− 0.01	0.02	71.6	4.7	356.3	363.0
AsH_3	66.6	0.02	0.04	64.0	8.8	218.4	212.7
SbH_3	57.8	− 0.01	0.03	55.7	5.8	184.5	184.3
OH	53.6	0.19	0.19	43.8	61.2	105.0	102.4
H_2O	55.8	0.12	0.18	49.6	62.2	223.6	221.6
H_2O_2	53.6	0.19	0.19	46.5	65.0	255.9	256.0
HS	73.2	0.07	0.07	68.1	17.1	85.2	87.1
H_2S	73.9	0.05	0.07	70.3	17.5	175.6	175.7
H_2Se	66.1	0.05	0.08	61.2	18.1	158.6	146.3
H_2Te''	63.0	0	0	63.4	0	126.8	125.9
HF	54.4	0.25	0.25	45.7	90.2	135.9	135.8
HCl	71.4	0.16	0.16	61.9	41.8	103.7	103.3
HBr	65.1	0.12	0.12	59.3	28.3	87.6	87.5
HI''	63.2	0.04	0.04	62.2	8.2	70.4	71.3

extent of the partial positive charge. This is an effect in addition to that of the weighting coefficients. Therefore the parameter to be used for positive hydrogen should be 104.2 $(1.00 - \delta_H)$ instead of 104.2 kpm. This correction produces striking improvement in the accuracy of the calculations. Table 8-3 summarizes the results for all simple binary compounds of hydrogen for which data are available, provided the hydrogen is positive, neutral, or only very slightly negative.

Some special problems associated with the C—H bond will be discussed more fully in Chapter 10. As will be seen from the data of Chapter 10, the above correction is also successful in all organic molecules that have been examined, for positive hydrogen in hydroxyl, carboxyl, amine, and amide, and sulfhydryl groups.

COMPOUNDS OF NEGATIVE HYDROGEN

Binary compounds of negatively charged hydrogen remain puzzling. Their chemical properties certainly conform to expectations, but the usual

TABLE 8-4

ATOMIZATION ENERGIES OF BINARY HYDRIDES

Compound	E_{AB}	δ_H	t_i	Atomization energy (kpm)	
				Calc.[a]	Exp.
LiH(g)	54.4	− 0.49	0.49	56.4	59.8
LiH(c)	54.4	− 0.49	0.49	132.9	112.1
NaH(g)	43.3	− 0.50	0.50	42.6	48.1
NaH(c)	43.3	− 0.50	0.50	99.2	91.7
KH(g)	37.1	− 0.60	0.60	37.8	43.5
KH(c)	37.1	− 0.60	0.60	89.1	88.0
RbH(g)	35.9	− 0.63	0.63	37.6	38.7
RbH(c)	35.9	− 0.63	0.63	88.7	92.0
CsH(g)	33.4	− 0.65	0.65	35.8	41.9
CsH(c)	33.4	− 0.65	0.65	84.0	82
BeH(g)	80.5	− 0.16	0.16	73.9	53
BH(g)	84.6	− 0.08	0.08	78.4	79.1
BH₃(g)	84.5	− 0.04	0.08	253.5	266.8
AlH(g)	68.6	− 0.19	0.19	65.7	68.1
GaH(g)	69.6	− 0.03	0.03	69.6	65.6
InH(g)	60.5	− 0.09	0.09	57.5	58.8
SiH(g)	74.7	− 0.09	0.09	73.2	74.7
SiH₄(g)	74.6	− 0.04	0.10	304.4	309.5
SnH(g)	62.3	− 0.06	0.06	59.9	74
SnH₄(g)	62.3	− 0.02	0.06	252.0	241.7

[a] All calculated as if $t_i = 0$. For solid hydrides, $n = 3$.

methods of bond energy calculation lead to energies that are invariably far too high. It is quite remarkable that if the energy of these molecules is assumed to be totally nonpolar covalent, regardless of the magnitude of the negative charge calculated for the hydrogen, the corresponding calculated energy is not only of the right order of magnitude but usually very close to the experimental value. This is demonstrated by a listing of all available data in Table 8-4.

An interesting related compound is lithium hydroxide, LiOH, for which the gaseous atomization energy has been determined to be 208 kpm. The partial charges in this molecule are Li, 0.90; O, -0.60; and H, -0.30. The Li—O bond length is reported as 1.82 Å, compared to the covalent radius sum of 2.06 Å. The total energy of the Li—O bond is calculated in the usual manner to be 145.3 kpm. If the O—H bond is assumed, in view of the highly negative hydrogen atom, to be equivalent to nonpolar, as in the hydrides of Table 8-4, and the bond length is the 0.98 reported for OH⁻ ion, the calculated nonpolar energy of the O—H bond is 63.2 kpm. The sum of the two bond energies is 208.5 kpm, essentially in agreement with the experimental value.

Obviously this is an area deserving closer study. All that can be done at the present writing is to emphasize that negative hydrogen with its very large percentage increase in electronic cloud must be extremely polarizable. Indeed, one may speculate from the results shown in Table 8-4 that the polarization is so great as to cancel effectively the bond polarity to the extent that polarity would contribute to bond energy. In general, it may be pointed out that the whole scheme of bond energy calculation as described and applied in this book depends on the concept of atoms retaining in combination, at least approximately, their individual character. It appears, and perhaps not too surprisingly, that in negative hydrogen the nuclear charge is too small to maintain the approximate condition of atomic individuality of the hydrogen. Hence the arbitrary separation of a covalent and an ionic contribution to the total bonding energy appears to become invalid for such compounds.

HYDRIDIC BRIDGING

Diborane

In the molecule of diborane, B_2H_6, two BH_2 groups are bridged by two hydrogen atoms. The bond lengths in the BH_2 groups are 1.19 Å and the B—H distance in the bridges is 1.33 Å. Each bridge appears to consist of a three-centered bond with two boron atoms and one hydrogen atom held together by two electrons. The partial charges are -0.04 on hydrogen and 0.13 on boron, giving $t_i = 0.09$. Calculated atomization energy was found to be much too low if bond polarity is ignored as for compounds of negative hydrogen in general. On the other hand, if polarity is considered, the covalent energy of the shorter type of bond is 73.7 and the ionic energy 25.1, totaling 98.8 kpm. The energy for four of these bonds is 395.2 kpm. The longer type of bond appears to be a half-bond. Its covalent energy is 65.9 and its ionic energy is 22.5, totaling 88.4 kpm. The four half-bonds are equivalent to two full bonds, totaling 176.8 kpm. Added to 395.2 this gives a total energy of atomization of diborane of 572.0 kpm, to be compared with the experimental value of 573.1 kpm (see Fig. 8-2).

A similar treatment of silane, SiH_4, in which the partial charge on hydrogen is also -0.04, and $t_i = 0.10$, gives an atomization energy more than 50 kpm too high, whereas ignoring the polarity leads to a calculation that is approximately correct. It appears necessary to assume that the presence of hydridic bridging in diborane somehow changes the condition of the B—H bonds so that their energy is appropriately calculated accounting for the polarity in the usual way. At present the difference is not understood.

Tetraborane

The tetraborane molecule, B_4H_{10}, should present no difficulty. It consists of two triangles of boron atoms having a common base and with the apices

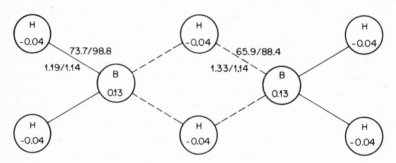

FIG. 8-2. Schematic representation of diborane. Atomization energy: 572.0, calc.; 573.1 kpm, exp.

forming a V-shape. The two base boron atoms are bonded together, and attached to each terminal boron atom by a hydridic bridge. One hydrogen atom normally bonded fills the fourth coordination space about each base boron, and two hydrogen atoms are normally attached by single bonds to the terminal boron atoms. The total number of valence electrons is twenty-two. If two electrons are assigned to each hydrogen whether bridging or not, two are then left to form a normal B—B bond between the base boron atoms. The partial charges are 0.12 on boron and -0.05 on hydrogen. The boron-boron bond energy is the normal homonuclear energy 68.4, reduced by the factor 0.88, since the positive charge on boron is 0.12, and corrected for the increase in length from 1.64 for the covalent radius sum to 1.71 Å. The calculated energy is 57.7 kpm.

The six normally bound hydrogen atoms are at a distance of 1.19 compared to 1.14 Å for the nonpolar covalent radius sum. The nonpolar energy value is 84.5. From the partial charges, $t_i = 0.08$. The value of E_c is then 74.5 and $E_i = 22.3$. Six such bonds have a total energy of 580.8 kpm. The bridging hydrogens are unsymmetrically placed, the distance to the terminal or apical boron atoms being 1.43, whereas to the base boron atoms it is 1.33 Å. The four half-bonds of shorter length have a total energy of 173.2, whereas the longer half-bonds have a total energy of 161.0 kpm. The sum for the molecule is then 972.7 kpm. But the National Bureau of Standards 1966 selection of heat of formation is $+15.8$ kpm, corresponding to an experimental atomization energy of 1043.2 kpm. This discrepancy cannot at present be accounted for. The calculated heat of formation is $+86.3$ kpm.

Pentaborane-9

This molecule, B_5H_9, has a square pyramidal shape. Four boron atoms form a square base 1.80 Å on a side. A fifth sits on top of the four with bond length to each, 1.69 Å. One hydrogen is bound by a normal covalent bond to each boron, and four more bridge the base boron atoms. The bonds to hydrogen being a little longer than in diborane or tetraborane, 1.22 instead of 1.19 Å, and 1.35 instead of 1.33 Å, one may suspect some involvement of the B—H bonds with the B—B bond system. However, as an approximation, one can assume the B—H bonds to be independent, and calculate their energies. The charges on boron and hydrogen are 0.11 and -0.06, making $t_i = 0.08$. For the normal B—H bonds, the energy is calculated to be covalent 72.7, ionic 21.8, total per bond 94.5, and 472.5 for the five bonds. The bridges may be assumed to constitute eight half-bonds, for which the total energy is calculated to be 341.6 kpm.

The complex system of boron atoms can be dealt with only approximately. However, if it is assumed that the B—H bonds have their normal supply of bonding electrons, then six are left out of the twenty-four total to participate in the boron–boron bonding. Under normal circumstances this is not enough for two-electron bonds between boron pairs. However, it does average 1.2 bonding electrons for each of the five boron atoms. Because of the special structure of the molecule, the apical boron atom being attached to four other boron atoms all on the same side of it, the bonding 1.2 electrons assigned to it may be thought of as serving fully and equally, or nearly so, all four bonds. If it is assumed that this boron can furnish, in effect, one electron to each of the four bonds, then with 1.2 electrons on each base boron atom there are enough for the equivalent of 4.8 normal single covalent bonds. Normally one would correct the boron homonuclear single covalent bond energy for the positive charge on the boron atoms, but this effect must be reduced to about half under the peculiar circumstances of crowding the bonds on one side of the apical boron. We can then roughly approximate the total boron–boron bond energy as $68.4 \times 0.95 \times 4.8 \times 1.64/1.69 = 302.1$ kpm. Added to the total B—H energy of 814.1 kpm, this gives a total estimated atomization energy of 1116.2 kpm. The experimental value is 1123.9 kpm. The approximation procedures appear to be reasonably successful. At least no large questions about the calculations appear necessary. The general concept of normal and half-bonds between boron and hydrogen does not seem seriously challenged here.

Decaborane

The structure of the $B_{10}H_{14}$ molecule is too complex for even a roughly approximate treatment of bond energies such as just described for B_5H_9. The B—H bond lengths are substantially higher than for diborane, the "normal" hydrogens being 1.25 and 1.28 Å from boron and the bridges being 1.34 and 1.40 Å long. The total system of nineteen boron–boron bonds, ten single B—H bonds, and eight half-bonds appears to require polycentric molecular orbital descriptions defying reasonably simple treatment by the bond energy methods discussed herein.

Aluminum Borohydride

This molecule AlB_3H_{12} is believed to consist of an aluminum attached to three BH_2 groups through a double hydridic bridge to each. The nonbridging and bridging B—H distances are reported as 1.21 and 1.28 Å. The aluminum to boron distance of 2.15 is probably more nearly correct than the reported Al—H distance of 2.0 which is inconsistent. A value of 1.76 Å may be determined from the Al—B distance and by analogy with diborane. The "normal" B—H bond energy is calculated to be covalent 73.2, ionic 21.9, total 95.1, times six bonds = 570.6 kpm. The energy of the six half-bonds is that of three full bonds (covalent 69.2 and ionic 20.7) totaling 269.7 kpm for the bridging to boron. The hydridic bridges to aluminum are equivalent to three full bonds of covalent energy 48.7 and ionic energy 39.6 kpm, totaling 88.3 kpm per bond or 264.9 kpm per molecule. Added to the boron–hydrogen energy, this gives 1105.2 for the atomization energy of AlB_3H_{12}. The experimental value, based on the recent National Bureau of Standards choice of +3 for the heat of formation, is 1103.7 kpm (see Fig. 8-3). The excellent agreement indicates that aluminum borohydride is another example of the importance of including the polar energy calculations for negative hydrogen where hydridic bridging occurs.

PROTONIC BRIDGING

As thoroughly discussed by others,[2] electrostatic energy is evidently an important component of protonic bridging (commonly called "hydrogen bonding"). It is interesting to see how the experimentally determined bridge energies compare with electrostatic energies calculated solely from the partial charges on the hydrogen on one molecule and the negative bridging

[2] G. C. Pimentel and A. L. McClellan, "The Hydrogen Bond." Freeman, San Francisco, 1960; S. H. Lin *in* "Physical Chemistry," (H. Eyring, D. Henderson and W. Jost, eds.) Vol. 5, p. 439, Academic Press, New York, 1970.

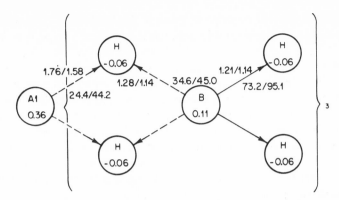

FIG. 8-3. Schematic representation of aluminum borohydride. Atomization energy: 1105.2 kpm, calc.; 1103.7 kpm, exp.

atom on the other molecule, disregarding all other possible interactions. None of the necessary distances appear to be known very accurately, nor have the experimental protonic bridge energies been evaluated exactly. In consideration of these limitations, the extent of agreement between the experimental and calculated bridge energies, shown in Table 8-5, is somewhat remarkable. At least the values are of an apparently correct order of

TABLE 8-5

ESTIMATION OF ELECTROSTATIC PROTONIC BRIDGE ENERGY IN GASES

Compound	δ_H	δ_X	R(X—H)	R(X...X)	R(XH...X)	$E_{calc.}$	$E_{exp.}$
NH$_3$	0.05	−0.16	(1.01)	3.10	(2.09)	1.3	3.8(g), 1.3(c)
H$_2$O	0.12	−0.25	(0.99)	2.76	(1.77)	5.5	5.0
HF	0.25	−0.25	(0.92)	2.55	(1.63)	12.7^a	6.8
CH$_3$OH	0.07	−0.29	(1.07)	(2.76)	(1.69)	5.6	6.0
HCOOH	0.16	−0.21	1.07	2.73	1.66	6.7	7.1
CH$_3$COOH	0.11	−0.26	(1.07)	2.76	(1.69)	5.6	7.0

^{a} If the linear zigzag polymers are planar, the bond angle of 120° permits alternate fluorine atoms to be close enough for about 5–6 kpm repulsion, reducing the calculated net energy by that amount.

magnitude. It may be noted that the calculated value for HF is nearly twice the experimental value. If, however, the staggered chain polymer molecules are planar, then the bond angle of about 120° is sufficient to bring alternate fluorine atoms significantly close together where they would introduce a repulsive force of several kpm. This would reduce the net protonic bridging energy as experimentally determined to about the observed value.

Surely protonic bridging is much more complex. There is no intention of suggesting otherwise by including such simple calculations.

116   8. APPLICATIONS TO INORGANIC MOLECULES

THE BIFLUORIDE ION

If one wishes to classify all bridging hydrogen as "hydrogen bonding," then the bifluoride ion, FHF^-, is indeed an example. However, as was pointed out some years ago,[3] this ion is definitely not an example of protonic bridging, for the partial charge on hydrogen is -0.04, and on each fluorine atom, -0.48. The FHF^- ion is known experimentally to be linear with the proton halfway between the two fluorine atoms. The F...F distance is 2.26 Å, making the H...F distance 1.13 Å. Several equivalent viewpoints are acceptable, but it is convenient here to consider the ion to consist of a hydride ion attached to two fluorine atoms. The species is then analogous to XeF_2, and can similarly be described as a three-center system involving four electrons. By the molecular orbital concept, the hydrogen orbital and one orbital from each fluorine atom form three molecular orbitals: one bonding, one nonbonding, and one antibonding. The four electrons, two from the hydride ion and one from each fluorine atom, occupy the bonding and the nonbonding molecular orbitals, creating in effect two half-bonds between fluorine and hydrogen. Following the principle that negative hydrogen behaves as though its bonding energy were completely nonpolar, we can now calculate the bond energy. The geometric mean of 37.8 for fluorine and 104.2 for hydrogen is 62.8.

$$E_c = \frac{62.8 \times 1.03 \times 0.5}{1.13} = 28.6 \text{ kpm}$$

The experimental value has been reported as about 27–28 kpm.

Boron Heterocycles

Because of the well-known similarity among isoelectronic species, compounds having B—N instead of C—C have attracted special interest. Prominent among these is the boron heterocycle, $B_3N_3H_6$, called "borazine" (sometimes "borazole"). This molecule bears a physical resemblance to that of benzene, having a six-membered ring isoelectronic with the carbon ring and one hydrogen atom attached to each atom of the ring. However, as shown in Fig. 8-4, the ring bears alternating charges, the bonds between boron and nitrogen being substantially polar. From bond energy calculations the B—N bonds appear to be represented as an average of B—N' and B—N", showing therefore only a relatively small increase in the stability over that expected for E' single bonds.

In all the other boron heterocycles for which gas phase data are available, the positive charge on boron is considerably larger. The bonds also are different, in being B—N" and B—O" bonds, as shown in Table 8-6. The

[3] R. T. Sanderson, *J. Chem. Phys.* **23**, 217 (1955).

only other nitrogen heterocycle is the borazine derivative in which the three hydrogen atoms on boron in borazine have been replaced by chlorine atoms. (Many others have been prepared but appropriate data are not available.)

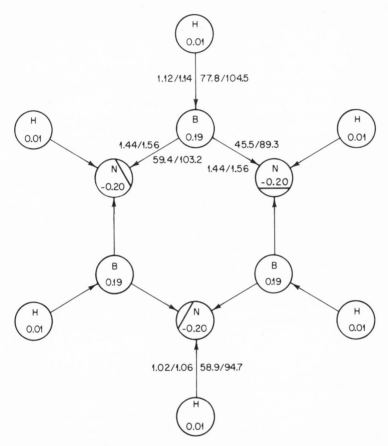

FIG. 8-4. Schematic representation of borazine (all ring bonds are equal). Atomization energy, 1175.1 kpm, calc.; 1177.4 kpm, exp.

All known compounds that have been investigated in which a single bond exists between boron and oxygen are of the B—O″ type as in the rings of the compounds listed in the table.

Molecular Addition Compounds

One area that has as yet been examined only briefly and superficially is that of the coordination bond. Gas phase data are available for a number of molecular addition compounds in the most recent National Bureau of Standards compilations. A study of these reveals some interesting possibilities for further exploration and is therefore reported here.

First, the structural change that accompanies the change in an M3 compound from coordination number three to four, from planar to tetrahedral,

TABLE 8-6

BOND ENERGIES IN SOME BORON HETEROCYCLES

Compound	Ring	δ_A	δ_B	E_c	E_i	Atomization energy (kpm)	
						Calc.	Exp.
$B_3N_3H_6$	B—N′	0.19	− 0.20	45.5	43.8	1181.1	1177.4
	B—N″	0.19	− 0.20	59.4	43.8	—	—
$B_3N_3H_3Cl_3$	B—N″	0.27	− 0.14	58.7	46.1	1218.6	1223.9
$B_3O_3(OH)_3$	B—O″	0.33	− 0.23	55.1	68.4	1454.1	1463.2
$B_3O_3H_2F$	B—O″	0.30	− 0.26	54.3	67.3	1084.1	1088.4
$B_3O_3HF_2$	B—O″	0.36	− 0.21	53.6	69.7	1141.2	1148.2
$B_3O_3F_3$	B—O″	0.42	− 0.16	54.4	70.8	1205.1	1205.8
$B_3O_3H_2Cl$	B—O″	0.28	− 0.27	55.1	65.0	1031.4	1029.6
$B_3O_3HCl_2$	B—O″	0.32	− 0.24	54.3	67.4	1040.1	1031.6
$B_3O_3Cl_3$	B—O″	0.36	− 0.21	54.4	70.8	1067.4	1061.0

has generally been thought to involve in itself an energy change. Such a change has not yet been revealed by the bond energy data to date.

Second, one might expect that a change in bond type would occur when the fourth orbital of boron, for example, becomes occupied by another donor atom. Thus far there is no evidence of such a change either. Addition compounds of BF_3, for instance, appear to involve B—F‴ bonds just as in BF_3 itself, expect that the bonds are appreciably longer and, therefore, correspondingly weaker.

A third point to be mentioned first attracted attention during the early development of the concepts of electronegativity equalization and partial charge. The question arose: Why should there be equalization of electronegativity throughout a molecular addition compound if the donor molecule is originally more electronegative than the acceptor molecule?[4]

The results of bond energy calculations for 1:1 molecular addition compounds are summarized in Table 8-7. This table lists the electronegativities of all the donor and acceptor molecules involved, the calculated and experimental atomization energies and their differences, and the quantity of charge transfer through the coordination bond, if electronegativity equalization occurs. The coordination bond is assumed to be of A′—B′

[4] R. T. Sanderson, "Chemical Periodicity," p. 66. Reinhold, New York, 1960.

TABLE 8-7

ATOMIZATION ENERGIES OF SOME 1: 1 MOLECULAR ADDITION COMPOUNDS

Compound	Electronegativity		Atomization energy (kpm)		Difference	Donor to acceptor charge transfer
	Acceptor	Donor	Calc.	Exp.		
$GaCl_3NH_3$	4.45	3.76	577.2	572.9	4.3	− 0.32
$BF_3P(CH_3)_3$	4.86	3.59	1571.7	1561.8	9.9	− 0.88
$BF_3N(CH_3)_3$	4.86	3.67	1598.6	1591.4	7.2	− 0.82
$BF_3O(C_2H_5)_2$	4.86	3.71	1815.7	1800.8	14.9	− 0.82
$BF_3O(CH_3)_2$	4.86	3.76	1259.5	1234.5	25.0	− 0.70
BH_3PF_3	3.38	5.05	709.7	626.8	82.9	+ 0.88
$BH_3N(CH_3)_3$	3.38	3.67	1485.5	1406.9	78.6	+ 0.23
$Ga(CH_3)_3S(CH_3)_2$	3.58	3.66	1893.1	1805.8	87.3	+ 0.09
$B(CH_3)_3P(CH_3)_3$	3.55	3.59	2335.5	2243.6	91.9	+ 0.06
$B(CH_3)_3N(CH_3)_3$	3.55	3.67	2353.1	2265.7	87.4	+ 0.19
$B(CH_3)_3NH(C_2H_5)_2$	3.55	3.66	2640.6	2551.5	89.1	+ 0.19
$B(CH_3)_3NH_3$	3.55	3.76	1531.4	1440.7	90.7	+ 0.16
$B(CH_3)_3NH_2CH_3$	3.55	3.71	1805.9	1714.7	91.2	+ 0.16
$BH_3S(C_2H_5)_2$	3.38	3.65	1686.1	1594.1	92.0	+ 0.22
H_3CO	3.38	4.45	644.3	548.3	96.0	+ 0.33
$BH_3S(CH_3)_2$	3.38	3.66	1125.5	1023.2	102.3	+ 0.20
$Al(CH_3)_3S(CH_3)_2$	3.47	3.66	1959.2	1826.6	132.6	+ 0.25

type in all compounds listed. The data perhaps conceal more than they reveal. They do indicate a remarkable distinction between those compounds in which the donor was originally less electronegative and those in which the donor was originally more electronegative than the acceptor. Briefly, the methods work quite well for the former class. Calculated atomization energies are much too high for all those addition compounds for which the equalnizatioel of ectronegativities would require the transfer of negative charge from the acceptor to the donor. It will further be noted that seven of the twelve compounds of this group show a reasonably constant difference between calculated and experimental values. This suggests a constant energy requirement for reversing the electron flow. Obviously much work is needed here. It is hoped that this brief discussion may stimulate further study.

Chapter 9

THE COORDINATED POLYMERIC MODEL OF "IONIC BONDING" IN NONMOLECULAR BINARY SOLIDS

Practically all chemists have been trained to believe that most binary compounds of metal with nonmetal, and especially the alkali metal halides, exist in the crystalline state as assemblages of ions. Indeed, this was probably the very first thing they learned about the chemical bond, for the concept is eminently logical and teachable. Mutual acquisition of "noble gas" structures by two atoms of active elements, through the direct process of transfer of electrons from one atom to the other, is an intuitively appealing idea because of both its simplicity and its effectiveness in explanation. The development of more sophisticated bonding theory has led to a general awareness, if not clear recognition, that "ionic bonding" probably represents an extreme condition relatively seldom actually realized, and that most real chemical bonds probably have some "covalent character." Nevertheless, the concept of ions in crystals is deeply imbedded in contemporary chemical thought. The average practicing chemist, if he should pause to reflect upon it, would probably be somewhat disturbed by the suggestion that an alternative model of the "ionic bond" might be preferable.

Since offering such a suggestion is exactly the purpose of this chapter, an objective review of the causes of belief in ionic bonding may prove to be an essential preliminary. Why is this concept so widely accepted?

Evidence of Crystalline Ions

Perhaps the chief arguments, following an earlier discussion,[1] can be summarized as follows:

[1] R. T. Sanderson, *J. Chem. Educ.* **44**, 516 (1967).

1. The stoichiometry of many binary metal–nonmetal solids is consistent with the mutual "stabilization" of atomic outer shells through transfer of electrons to form ions, initially incomplete electron groups being either removed or completed.

2. All structural studies show these compounds to be nonmolecular, each internal atom being surrounded by several atoms of the other kind, usually at equal distances revealing no "preference" for one close neighbor over any other.

3. Such solids commonly dissolve in polar solvents, especially water, giving solutions that by their conductance indicate the abundance of ions. The charges on the ions are confirmed by conformance with Faraday's law of electrolytic equivalence.

4. In the fused state, such substances readily conduct an electric current with simultaneous deposit of elements at the electrodes which appear to result from the discharge of their ions, again conforming to Faraday's law.

5. Ionic radii determined by reasonable procedures are found to be nearly invariant and additive to give approximately the correct internuclear distances in the crystals.

6. The Born-Mayer equation,[2] which is based directly on the ionic model, permits quite accurate calculation of the crystal energy for a number of salts, principally the alkali halides.

7. Detailed X-ray diffraction studies revealing the electronic density pattern throughout the crystal have been interpreted as implying very close to the ionic number of electrons around each nucleus in several compounds.

If other compelling arguments exist, they are omitted inadvertently, for the above list is intended to be reasonably comprehensive.

Let us consider these arguments in the same order, one by one.

1. The first argument is correct to the extent that atoms of different elements do tend to combine in relative proportions consistent with ion formation. For example, sodium reacts with chlorine 1:1, as it would if one electron from each sodium atom were transferred to a chlorine atom. This process would leave sodium deficient by one electron and isoelectronic with the M8 element neon. Chlorine with an excess of one electron per atom would then be isoelectronic with the M8 element argon. It is also true that sodium atoms and chlorine atoms each have one half-filled orbital in their outermost quantum shell, permitting them to join together 1:1 by forming a single covalent bond. Therefore ionicity is not established by the mere fact of stoichiometry.

Furthermore the alleged special stability of structures isoelectronic with

[2] M. Born and J. E. Mayer, *Z. Phys.* **75,** 1 (1932); T. C. Waddington, *in* "Advances in Inorganic Chemistry and Radiochemistry" (H. J. Emeleus and A. G. Sharpe, eds.) Vol. 5. Academic Press, New York, 1959.

the M8 elements is a myth. All ions that are isoelectronic with M8 elements possess far greater chemical reactivity because their nuclear charges are not like those of the M8 elements. Also, all cations and some simple anions can be formed from the corresponding atoms only endothermically. The energy of separating a compound into its atoms is always less than the energy of separating it into ions.

2. The second argument is consistent with the ionic model but hardly proves its validity. A metallic crystal is similarly nonmolecular. So is a covalent polymer such as silicon. The absence of identifiable molecules in the solid state does not prove that the atoms are present as ions.

3. The third kind of evidence is consistent with the ionic model also, but again does not exclude other possibilities. One needs only to think of hydrogen chloride and hydrochloric acid to recognize that the formation of ions in solution does not prove their existence in the pure compound.

4. With regard to the fourth argument, it is tempting to think of the conductance of fused salts and the electrodeposition of their component elements as convincing evidence that they consist of ions in the melt and probably therefore in the crystal. Yet what do we really know about the species that constitute the melt? A little salt will permit conductance of water with the liberation of hydrogen and oxygen at the electrodes. Does this prove that water consists of ions? Moreover, high temperatures are required to melt most salts. If ions were present at high temperature in the liquid state, does this prove their existence in the crystal at ordinary temperature?

5. The fifth point pertaining to ionic radii has never been very convincing because for many compounds the nonpolar covalent radius sums and the ionic radius sums are nearly equal. The point becomes even less convincing in the light of careful X-ray examination of the way in which the electronic density varies in a crystal along a straight line from the nucleus of a metal atom to the nucleus of an adjacent nonmetal atom. If ions as such have any individual identity within the crystal, one might expect a density minimum where the two ions meet, owing to mutual repulsion between their electronic clouds. A density minimum is indeed observed. It is not easy to imagine any possible interpretation of such a minimum other than as representing the point of junction between the ions. But to the thoughtful advocate of the ionic model the internuclear electronic density minimum in a crystal has two disturbing features. First, it is never zero. There is a continuum of electronic density from one nucleus to the next. Second, and even more disconcerting, is the fact that the minimum never occurs[3] where it would

[3] K. B. Harvey and G. B. Porter, "Introduction to Physical Inorganic Chemistry," Addison-Wesley, Reading, Massachusetts, 1963; J. C. Slater, "Quantum Theory of Molecules and Solids," Vol. 2, p. 107. McGraw-Hill, New York, 1965.

be expected if ionic radii were the determining factor. The X-ray "radius" of a nonmetal atom in a crystalline binary compound is always appreciably smaller than the conventional ionic radius, but larger than the nonpolar covalent radius. The X-ray radius of a metal atom in such a compound is always appreciably larger than the conventional ionic radius but smaller than the nonpolar covalent radius. Since radii are expected to increase with negative charge and decrease with positive charge, these observations, qualitatively at least, are exactly what would be expected of partially charged atoms. A comparison of partial charge, nonpolar covalent radius, X-ray radius, and ionic radius is permitted by the data of Table 9-1.

TABLE 9-1

COMPARISON OF X-RAY RADII WITH COVALENT AND IONIC RADII

| Compound | Atom | Charge | Radius (Å) | | | Ion |
			Nonpolar covalent	X-ray	Ionic	
CaF_2	F	-0.47	0.71	1.10	1.33	F^-
LiF	F	-0.74	0.71	1.09	1.33	F^-
CuCl	Cl	-0.29	0.99	1.25	1.81	Cl^-
NaCl	Cl	-0.67	0.99	1.64	1.81	Cl^-
KCl	Cl	-0.76	0.99	1.70	1.81	Cl^-
CuBr	Br	-0.25	1.14	1.36	1.96	Br^-
MgO	O	-0.50	0.72	1.09	1.45	O^{2-}
LiF	Li	0.74	1.34	0.92	0.68	Li^+
NaCl	Na	0.67	1.54	1.18	0.98	Na^+
KCl	K	0.76	1.96	1.45	1.33	K^+
MgO	Mg	0.50	1.38	1.02	0.65	Mg^{2+}
CaF_2	Ca	0.94	1.74	1.26	0.94	Ca^{2+}
CuCl	Cu	0.29	1.35	1.10	0.96	Cu^+
CuBr	Cu	0.25	1.35	1.10	0.96	Cu^+

It seems evident from these data that if the conventional ionic radii are indeed accurate measures of the size of ions, then ions cannot be present in the crystals of compounds generally thought to contain them. Alternatively, if the conventional ionic radii are not accurate measures of the sizes of ions, point 5 becomes meaningless as evidence in support of the ionic model.

6. The Born-Mayer equation does give the crystal energies of the alkali halides with striking success. However, it becomes decreasingly satisfactory when applied to other salts, and fails for those compounds generally recognized as having considerable "covalent character."

7. Finally, the X-ray evidence for the electron population around the separate nuclei is open to some question because it is not sufficiently accurate to distinguish between complete and partial ionicity and because of the appreciable electronic density observed even at the minimum. Furthermore, J. C. Slater has pointed out[4] that even an exact knowledge of the electron density distribution within a crystal could not serve to establish the degree of ionicity.

In summary, the evidence for the existence of ions in solids seems reasonable, appealing, and persuasive. It is also indirect, incomplete, ambiguous, and inconclusive.

Is the Ionic Model Reasonable?

To the degree that it may be possible, let us put aside the prejudices of past training and try to consider objectively whether the ionic model of binary solids makes good sense.

First, what is the true nature of ions? According to the ionic model, they are conveniently treated as though they were more or less rigid spheres having positive or negative charges residing at their centers. Purely through the electrostatic interaction of these charges, these spheres become arranged within the crystal in such a way that each positive charge is surrounded as closely as possible by negative charges, and as far removed as possible from other positive charges. Similarly, each negative charge is surrounded as closely as possible by positive charges, and as far removed as possible from other negative charges. A compromise is necessary, resulting in the presence of unlike charges as closest neighbors but like charges just beyond them. Therefore the potential energy of each "ion" must be integrated over the entire crystal as an elaborate series of nearby attractions, repulsions a little farther away, additional attractions a little farther yet, and so on. The summation of this series is the Madelung constant, a dimensionless constant by which the electrostatic energy between two adjacent opposite charges must be multiplied to represent the total coulombic energy of the crystal: $E = 332 \, M \, z^+ \, z^- \, e^2/R$. But the rigid sphere concept of the ions must be modified in recognition of the fact that they are actually electronic clouds. Equilibrium among the various forces can only be reached when the net electrostatic attractions among the ionic charges are balanced by the repulsions among the adjacent electronic spheres which prevent mutual penetration. Born recognized this necessity by including a "repulsion coefficient," which was called k in Chapter 2 and which will be discussed later in this chapter.

[4] J. C. Slater, "Quantum Theory of Molecular Solids," Vol. 2, p. 107. McGraw-Hill, New York, 1965.

The above question (What is the true nature of ions?) is really intended, then, to ask: What is wrong with representing ions in the manner described? In the first place, the simple picture of alternating negatively charged particles and positively charged particles is not realistic enough. Nor is it sufficient to recognize the nature of the spheres as electronic clouds and account for their mutual repulsion. For an ion individually has special characteristics not dealt with in the Born-Mayer treatment.[5] First, no ion whether positive or negative is a point charge. All simple ions consist of a positively charged nucleus imbedded in an electronic cloud. In a cation the total negative charge on the cloud is inadequate to balance the nuclear charge. In an anion, the total negative charge on the cloud is more than enough to balance the nuclear charge. The attraction between a cation and an anion is really an attraction between the nuclear charge of the cation and the electronic cloud of the anion. But more than that is involved.

Every simple cation has an electronic structure featuring an outermost quantum shell of empty orbitals. The fact that they are empty does not mean they lack ability to hold electrons. Indeed every cation is capable of acquiring one or more electrons with the *evolution* of energy. Now, in coordination chemistry: What is an electron pair acceptor? Always it consists of an atom, the positive nuclear charge of which is not completely shielded from one or more outer shell orbitals that are vacant. That is exactly the condition of a cation. In other words, all simple cations possess the requisites to act as electron pair acceptors in the formation of coordination bonds.

Every simple anion has an electronic structure featuring an outermost quantum shell of filled orbitals, or electron pairs. Every simple anion also possesses a surplus of negative charge, which makes the outer pairs the more available to interact with a potential electron pair acceptor through its outer shell low energy vacant orbitals. In coordination chemistry: What is an electron pair donor? It is any atom bearing usually both an excess of electrons and one or more outer electron pairs. In other words, all simple anions are potential electron pair donors.

What is wrong with representing ions according to the ionic model of binary solids is that no account whatever is taken of the inherent donor qualities of the anion or the inherent acceptor qualities of the cation. If the ions were completely separated, this would be defensible. But the ions are not separated—they are in direct contact with one another. The question the staunch defender of the ionic model must answer successfully is this: How can there exist within a crystalline solid or anywhere else, an atom that is a potential electron pair donor, completely surrounded by and in close contact with other atoms that are potential electron pair acceptors, and also an atom that is a potential electron pair acceptor, surrounded by

[5] R. T. Sanderson, in *Advan. Chem. Ser.* No. 62, p. 187 (1967).

and in close contact with other atoms that are potential electron pair donors, and all without any donor–acceptor interaction? Is this situation reasonable?

Much impressive rationalization has appeared in the literature of chemistry concerning the relative influence of ionization energy, electron affinity, and coulomb energy of the ionic lattice. For example, it is recognized that the formation of a magnesium ion by removing two electrons from an atom of magnesium requires the sum of the first two ionization energies, or 526 kpm. This process must be preceded by the atomization of the magnesium metal, which requires a reported 36 kpm. Furthermore, various estimates have led to the conclusion that although a single electron is acquired by an oxygen atom exothermically, the acquisition of two electrons to form the gaseous oxide ion, O^{2-}, requires about 170 kpm. And first the O_2 molecule must be separated into atoms, at a cost of 59.6 kpm of atoms. The reason given for the well-known highly energetic combustion of magnesium metal to the crystalline solid MgO is that despite the 792 kcal deficit just enumerated, the energy to be derived from collecting gaseous Mg^{2+} ions together with an equal number of gaseous O^{2-} ions is more than enough to compensate. The experimental standard heat of formation of MgO(c) is indeed -143.8 kpm.

There is good reason to believe that the above reasoning is quite irrelevant. Ionization energy is the energy required to separate electrons from their cation (from their atom, leaving a cation) *completely*. Electron affinity is the energy released when a neutral atom acquires an electron *completely*. Examination of Fig. 9-1 will reveal why the arguments based on ionization energies and electron affinities are irrelevant. Nothing approaching complete separation of electrons occurs. Magnesium oxide is a substance in which the internuclear distance as measured experimentally sheds no light whatever on the nature of the bonding, since the ionic radius sum and the covalent radius sum are almost exactly the same. When we try to apportion the valence electrons between the magnesium atom and the oxygen atom, what we are really doing is deciding on the title to part of the region these electrons occupy. Except for their very important penetration toward the nucleus of one atom or the other, they occupy almost the same external region whether they belong to the magnesium or to the oxygen.

It is easier to distinguish the ownership of an electron in terms of what the electron does rather than where it is. If the valence electrons in a crystal penetrate exclusively toward one kind of nucleus and not at all toward the other, the crystal may be regarded as completely ionic. If, however, the valence electrons penetrate toward both kinds of nuclei, however unevenly, a covalent nature is thus introduced into the bonding. In the absence of experimental criteria, it is necessary to rely on reason, the infallibility of which unfortunately can never be trusted. However, if magnesium oxide is accepted as completely ionic by the above definition, the following implication seems inescapable.

Magnesium oxide is a crystalline solid in which each magnesium ion, which could acquire two electrons with the release of 526 kpm and possesses at least four stable orbitals which could accommodate them, is octahedrally surrounded by six oxide ions in direct contact with it. Each of these oxide ions could lose one electron with the release of 210 kpm, and two electrons with the release of 170 kpm. Each oxide ion, which has four readily available outer electron pairs, is likewise octahedrally surrounded by six magnesium ions in direct contact with it. This proximity has no effect whatever on the

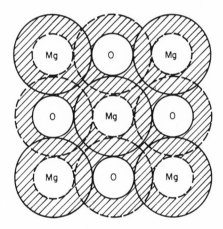

FIG. 9-1. Covalent and ionic magnesium oxide. Solid lines represent nonpolar covalent radii. Broken lines represent ionic radii. Striated area represents region occupied by valence electrons.

behavior of the electrons, except for the repulsions between the adjacent clouds. This implication not only seems inescapable. It also seems unacceptable.

Other problems also arise when we try to apply logic and reason to the ionic model. For example: What are the chemical properties of cations and anions? If a magnesium ion could capture two electrons with the release of 526 kpm, why aren't magnesium salts strong oxidizing agents? If an oxide ion could get rid of one electron with the evolution of 210 kpm, why aren't "ionic" oxides powerful reducing agents? If a magnesium ion is a good enough electron pair acceptor to coordinate with water molecules so exothermically as to release about 450 kpm, when the partial charge on oxygen in water is only -0.25, then why would it fail to coordinate with a donor oxygen of partial charge -2.00? And especially when the oxygen is known to be so efficient a donor that it separates a proton from a hydroxide ion by combining with it:

$$O^{2-} + H_2O \rightarrow 2\,OH^-$$

There is no need to confine this discussion to oxide ions. They are only one example, selected on the basis of the fact that oxide ions remain a common species in the thought and writings of students of solid oxides and conspicuous in applications of the ionic model of nonmolecular solids.

Then what of the successes of the Born-Mayer treatment? This approach has been unquestionably successful in the treatment of what seem to be the most nearly ionic of compounds, and especially of the alkali metal halides. But the alkali metal cations are the poorest of all cations in their electron pair accepting ability. It appears that where coordination between cation and anion in the crystal would be very weak, any errors in neglecting this interaction either cancel one another or are negligibly small. With improving donor and acceptor qualities, the Born-Mayer treatment becomes less successful and various modifications must be introduced to improve the agreement between calculated and experimental quantities.[6] So, with full recognition that successful application of any given physical model does not necessarily prove its validity in representing reality, let us examine a model that at least appears more generally applicable and less objectionable than the ionic model.

The Coordinated Polymeric Model

One may approach the state of a binary nonmolecular solid in two different ways, both imaginary and both leading to the observed result. One way is to begin with gaseous cations and gaseous anions in the appropriate stoichiometric quantities and allow them to come together. They will collect in the orderly array expected of a crystalline solid in such a way as to minimize the potential energy. The physical form of this crystal will be indistinguishable from that which is observed in nature and usually interpreted in terms of the ionic model. The cations and anions, however, will interact further as the result of the electron pair accepting power of the outer vacancies of the cations, together with the electron pair donating power of the negatively charged anions. The outer electron pairs of the anions, all of them and not merely what are conventionally considered the "valence electrons," will to a certain degree begin to penetrate toward the nuclei of

    [6] T. C. Waddington, *in* "Advances in Inorganic Chemistry and Radiochemistry" (H. J. Emeleus and A. G. Sharpe, eds.), Vol. I. Academic Press, New York, 1959; M. F. C. Ladd and W. H. Lee, *in* "Progress in Solid State Chemistry" (H. Reiss, ed.), Vol. I. Macmillan, New York, 1964.

the cations that surround them. This will diminish the electron-attracting power of the cations and increase the effective nuclear charge of the anions until the outer shell electrons on the average are attracted equally to both kinds of nucleus. The final equilibrium condition of the crystal can be described adequately as a coordinated polymeric condition in which all the atoms are held together through highly polar covalent bonding. The electron pairs are not necessarily localized, however, as in the conventional covalent bond, but participate equally in the bonding among all closest neighbors. The degree of polarity will be that determined by the principle of electronegativity equalization.

The potential energy of such a system is not as low as that which would correspond to a collection of integrally charged particles. One may well inquire then why, when such charged particles are brought together, the system would spontaneously depart from the condition of minimum potential energy to a higher energy level. The answer is that the system does not go to a condition of more than minimum potential. The confusion comes from the fact that the charged particles are not simple charged spheres but ions. What produces the most stable arrangement is not just the way the *ions* become arranged, but the way in which the *electrons* become arranged around the positive charges that lie at the center of both cations and anions. In other words, the coordinated polymeric model implies that the ionic model does not represent the most stable possible condition of the atoms. Presumably, reality does. Therefore a model that comes closer to reality is preferable.

The second way of approaching the state of a binary nonmolecular solid is to imagine beginning with the corresponding gaseous atoms. To simplify the picture, let us imagine these to be univalent atoms of both metal and nonmetal. The meeting of a metal atom with a nonmetal atom results in the formation of a diatomic molecule held together by a single covalent bond. Because of the initial electronegativity difference, the two bonding electrons spend more than half time more closely associated with the nonmetal nucleus than with the metal nucleus. This imparts a partial negative charge on the nonmetal atom and leaves a partial positive charge on the metal atom. In the sense that all half-filled orbitals of each atom are fully utilized in the covalent bond between them, the bonding capacity of each atom is fully satisfied. However, if the substance were to remain as individual diatomic molecules, there would be no effective utilization either of the vacant orbitals remaining on the metal atom or the lone pairs remaining on the nonmetal atom. The positive charge on the metal atom would enhance the availability of the vacant orbitals. The negative charge on the nonmetal atom would enhance the availability of the lone pairs for sharing. In conformance with an important rule of bonding to which only the boron halides are known exceptions, all molecules possessing both available orbitals and available electron pairs on separate atoms tend to condense further to make fuller use of the extra bonding capacity which these features provide. The diatomic molecules, initially attracted to one another through their negative and positive partial charges, condense to a nonmolecular solid. Within it the identity of all original molecules is lost. The most stable condition is reached when each interior atom is surrounded by atoms of the other kind. In the absence of special causes of distortion, these neighbors are all located in equivalent positions.

In this condition, the solid polymer is indistinguishable from the one formed by the condensation of the ions. Before considering the special virtues of this model, as well as the necessary precautions to the acceptance of any model, let us review in detail the method of atomization energy calculation described in Chapter 2.

THE COVALENT CONTRIBUTION

The contribution made to the total atomization energy by the coordination occurring between metal and nonmetal in the nonmolecular solid is calculated exactly as is the nonpolar covalent energy in a bond in a molecule, but with the addition of the factor n. That is, the nonpolar covalent energy of a nonmolecular solid is the geometric mean of the homonuclear single covalent bond energy parameters, corrected for any difference between observed internuclear distance and covalent radius sum, and multiplied by a small whole number n. The necessity for n arises from the fact that not merely the conventional valence electrons, but all the outer shell electrons of the formula unit may participate in the coordination bonding, and usually do. The number of "valence" electrons per formula unit is normally eight, instead of two, or some other lower number.

It is convenient, but not wholly satisfactory, to define the factor n as the equivalent number of two electron bonds formed per formula unit of the compound. Thus n should equal 4 regardless of whether the coordination number is 4, 6, or 8, the most commonly observed in the solid state. For example, in KCl, each atom is surrounded by six of the other kind. Here eight electrons are called upon to form six bonds, making each bond of order 0.67. But six bonds of 0.67 bond order are equivalent to four 2-electron bonds, and $n = 4$. Another example is the cesium chloride structure which involves 8:8 coordination. With only eight electrons to form four bonds, the bond order must be 0.5. Eight bonds of bond order 0.5 are the equivalent of four electron-pair bonds, so $n = 4$.

For reasons not yet fully understood, this procedure is not always satisfactory. For example, $n = 4$ applies very well to the cesium chloride structures of TlCl, TlBr, and TlI, but for CsCl, CsBr, and CsCl, n appears better evaluated as 6. For most binary compounds of the smaller metal atoms, lithium, sodium, beryllium, magnesium, zinc, cadmium, and probably aluminum, the correct value of n appears to be 3. Appropriate values of n are listed in the tables of atomization energy data. Further discussion may be included where useful in examining the specific applications later in this chapter.

In keeping with the concept of coordination polymerization, it will be shown later (Table 9-3) that although the nonpolar covalent contribution to the atomization energy of highly polar crystals is usually relatively small, it does constitute a significantly higher percentage of the total energy than in the gaseous state.

THE IONIC CONTRIBUTION

For the nonmolecular solid, the ionic energy differs from the similar energy for a polar covalent bond in a molecule in that it is also a coulomb energy defined by charge squared over distance, but must be multiplied by the Madelung constant and the repulsion coefficient:

$$t_i E_i = \frac{332 \, t_i \, M \, z^+ \, z^- \, e^2 \, R_c \, k}{R_0} \tag{9-1}$$

Wherever, as for some compounds, the structure remains in some doubt, so, also, does the Madelung constant. For most of the simple binary solids, the Madelung constant is known with much greater accuracy than is needed for the atomization energy calculation.

Where the structure remains in some doubt, the internuclear distances are also in some doubt. Even where structures appear to have been identified satisfactorily, significant discrepancies in bond length are occasionally reported in the literature Naturally the accuracy of atomization energy calculations is thus jeopardized.

Perhaps the least certain parameter needed for this calculation is the repulsion coefficient. What has been represented as k in the equation above was first represented as the factor $(1 - 1/n)$. The value of n can be determined experimentally by measurement of the compressibility, c, of a crystal by application of the equation,

$$c = \frac{18 R_0^4}{M e^2 (n - 1)} \tag{9-2}$$

One of various proposed modifications is that of Baughan,[7] whose method has been used for most of the values of k adopted here. He set n in the above expression $(1 - 1/n)$ equal to an empirical constant, g_0, times the bond length. He evaluated g_0 for a number of structure types, such as CaF_2 for

[7] E. C. Baughan, *Trans. Faraday Soc.* **55,** 736 (1959).

fluorite, for which he gives $g_0 = 3.20$. Since the ionic contributions are usually major parts of the total atomization energy, a small change in the assigned value of k could cause a change of a few kilocalories in the total energy. The extent of agreement obtained between calculated and experimental atomization energies of the more than one hundred nonmolecular compounds listed below suggests that errors in k have not in general been serious, although the possibility must not be overlooked.

Closely related to the problem of evaluating the repulsion coefficient, perhaps, is a difficulty encountered in the calculations for all nonmolecular binary compounds of polyvalent nonmetals. For all these, oxides, sulfides, and selenides, the calculated atomization energies are substantially too high. They are brought into reasonably good agreement with experimental values, however, if the ionic contributions are multiplied by a factor which is 0.63 for oxides and 0.57 for sulfides, and about 0.42 for selenides. The only suggestion, purely speculative, that has come to mind thus far to account for these factors is that possibly evaluations of k from compressibility depend too strongly on assuming the actual existence of ions in the crystal. For example, through application of the principle of electronegativity equalization one may conclude that CaO is only 28% ionic. Presumably then calcium oxide as it exists would be much less compressible than if it really contained oxide ions, O^{2-}, which should be a highly polarizable species by virtue of the large excess of electrons. From Eq. (9-2) we see that a larger compressibility would correspond to a smaller value of n. In turn this would correspond to a larger value of $1/n$. In turn this would correspond to a smaller value of $(1 - 1/n)$. Perhaps the value for k would be only 0.63 times as large for a truly ionic calcium oxide as it is for actual calcium oxide. In such compounds, by far the greater part of the compressibility would be expected to result from the highly polarizable anion. This would account for a nearly constant correction factor for all oxides. Furthermore, the larger sulfide and selenide ions would be still more polarizable and compressible, which would be consistent with the smaller correction factors found for their compounds.

As will be amply demonstrated, the consistency with which these "correction" factors apply suggests that whatever their origin, they do represent a real effect on the atomization energies of the chalcides.

THE SOLID ALKALI METAL COMPOUNDS

In these, as in all nonmolecular solids, only the single prime covalent bond energy parameters are applicable. The atomization energy calculations for these salts are summarized in Table 9-2. The difficulties encountered with the gaseous halides are completely absent here. The quality of the agreement between calculated and experimental atomization energies demonstrates beyond reasonable doubt the effectiveness of the coordinated polymeric model of these compounds. The fact that the ionic model suits them equally well should serve to emphasize that success within a narrow range of compound types does not necessarily prove the superiority of

TABLE 9-2

ATOMIZATION ENERGIES OF SOLID ALKALI METAL COMPOUNDS

Compound	Structure	R_o	n	M	k	t_i	E_c	E_i	Atomization energy (kpm)	
									Calc.	Exp.
LiF	NaCl	2.01	3	1.75	0.85	0.74	26.0	181.8	207.8	203.8
NaF	NaCl	2.31	3	1.75	0.87	0.75	19.1	164.1	183.2	181.8
KF	NaCl	2.67	4	1.75	0.88	0.84	14.3	160.9	175.2	175.8
RbF	NaCl	2.82	4	1.75	0.89	0.86	12.3	157.7	170.0	169.8
CsF	NaCl	3.00	4	1.75	0.90	0.90	8.2	156.9	165.1	164.5
LiCl	NaCl	2.57	3	1.75	0.88	0.65	38.6	129.3	167.9	165.2
NaCl	NaCl	2.82	3	1.75	0.89	0.67	28.8	122.9	151.7	153.2
KCl	NaCl	3.15	4	1.75	0.90	0.76	24.9	126.2	151.1	154.6
RbCl	NaCl	3.29	4	1.75	0.91	0.78	22.7	125.3	148.0	151.6
CsCl	CsCl	3.51	6	1.76	0.90	0.81	27.1	121.4	148.5	151.3
LiBr	NaCl	2.75	3	1.75	0.88	0.61	38.1	113.4	141.5	149.0
NaBr	NaCl	2.98	3	1.75	0.90	0.62	29.5	108.8	138.3	138.6
KBr	NaCl	3.30	4	1.75	0.91	0.71	26.8	113.8	140.6	138.6
RbBr	NaCl	3.43	4	1.75	0.91	0.73	24.8	112.5	137.3	139.3
CsBr	CsCl	3.71	6	1.76	0.91	0.77	28.8	110.4	139.2	139.7
LiI	NaCl	3.00	3	1.75	0.90	0.53	40.2	92.4	132.6	128.5
NaI	NaCl	3.24	3	1.75	0.90	0.54	31.2	87.2	118.4	120.2
KI	NaCl	3.53	4	1.75	0.91	0.63	31.0	94.4	125.4	125.1
RbI	NaCl	3.66	4	1.75	0.92	0.65	28.1	94.9	123.0	123.6
CsI	CsCl	3.95	6	1.76	0.92	0.69	33.9	93.9	127.8	124.7
Li_2O	CaF_2	2.00	6	5.04	0.83	0.40	115.3	175.0	290.3	279.5
Na_2O	CaF_2	2.41	4	5.04	0.85	0.40	56.1	148.7	204.8	210.8
K_2O	CaF_2	2.79	4	5.04	0.88	0.42	47.3	139.6	186.9	189.2
Rb_2O	CaF_2	2.92	4	5.04	0.88	0.43	46.1	136.9	183.0	177.8
Cs_2O	CdI_2	2.86	4	4.38	0.89	0.45	45.1	128.3	173.4	173.0
Li_2S	CaF_2	2.48	6	5.04	0.86	0.33	152.4	109.1	261.5	250.8
Na_2S	CaF_2	2.82	6	5.04	0.86	0.34	114.1	98.9	213.0	207.4
K_2S	CaF_2	3.20	6	5.04	0.89	0.37	101.2	98.1	199.3	211.8
Rb_2S	CaF_2	3.32	6	5.04	0.90	0.37	95.2	95.7	190.9	189.0
Cs_2S	CaF_2	3.48	6	5.04	0.90	0.41	83.8	101.1	184.9	185.1

either model. Neither does it establish either as a better representation of reality. From this viewpoint, both models could be in error and still give the good results observed for these compounds.

One may observe in the change from the rocksalt to the cesium chloride an accompanying change in the value of n from 4 to 6. Since there are only three examples here, it is not certain whether this change is justifiable. The thallium (I) halides except the fluoride also have the cesium chloride structure, as will be seen, and for them $n = 4$. Here, however, there are lone pairs on the thallium atoms, one on each, so the situation is not the same as for cesium.

Another puzzling feature of these calculations is the use of $n = 6$ for the antifluorite structure of the oxides and sulfides, whereas $n = 4$ in the fluorite

TABLE 9-3

CONDENSATION ENERGY OF ALKALI HALIDE GAS MOLECULES AND COMPARISON OF COVALENT CONTRIBUTION TO ATOMIZATION ENERGY BETWEEN GAS AND CRYSTAL

Compound	Condensation energy (kpm)	% Covalent energy	
		Gas	Solid
LiF	67.0	8.2	12.7
LiCl	50.9	14.2	23.3
LiBr	47.1	15.8	25.6
LiI	42.8	19.6	31.3
NaF	67.0	6.9	10.5
NaCl	54.2	11.6	18.8
NaBr	49.8	13.2	21.3
NaI	48.0	17.2	26.0
KF	57.7	3.6	8.1
KCl	53.0	7.3	16.1
KBr	51.2	8.7	18.8
KI	47.3	11.2	24.8
RbF	54.1	3.6	7.2
RbCl	51.1	6.6	17.8
RbBr	48.9	8.0	17.8
RbI	46.8	10.5	22.8
CsF	48.3	2.2	5.0
CsCl	50.3	5.5	17.9
CsBr	48.7	6.4	20.6
CsI	49.7	9.1	27.2

structured compounds. The value of 4 does appear appropriate for the oxides but not the sulfides of sodium, potassium, and rubidium. Cesium oxide has an altogether different structure, that of cadmium iodide.

In Table 9-3 are listed the energies of condensation from the alkali halide gas molecules to the crystalline state. They are very appreciable, as would be expected from the enhanced electrostatic interaction possible in the condensed state. However, other data in the table show that the condensation energy is not solely the result of increased electrostatic energy. These data represent the percentage of the total bonding energy resulting from non-polar covalent contributions. The experimental total energies were used to calculate these percentages, but of course the covalent contributions needed were the calculated ones. For all the alkali halides, the covalent contribution to the gaseous bond energy averages 9.5% of the total energy. The covalent contribution to the solid atomization energy averages 18.7%. Since the total bonding energy on the average is about 50% higher in the solid than in the gas, this represents a very substantial increase in energy resulting from coordination between metal and nonmetal in the solid.

SOLID M2 COMPOUNDS

Atomization energies of M2 halides, oxides, and sulfides are calculated with data summarized in Table 9-4. These calculations were not without problems in assignment of n. Although the value of $n = 3$ is appropriate for compounds of the smaller metal atoms in general, two exceptions appear, in BeF_2 and BeO where $n = 4$. The covalent energy contributions in both these compounds are large enough that this difference is quite significant and real. Otherwise the calculations are consistent except for calcium chloride and bromide. These apparently have highly deformed structures such that their atomization energies cannot be easily calculated. A value of $n = 3.5$ comes reasonably close to representing them but no explanation is available, and the results are highly tentative.

It is interesting to note that among the dihalides of such seemingly similar elements as magnesium, calcium, strontium, and barium such a remarkable variety of crystal structures occur. One may observe a tendency for the most ionic of these compounds to exhibit the fluorite structure, or the $PbCl_2$ structure. Halogen partial charges of about -0.40 or lower tend to correspond to modifications of the rutile structure with layer structures appearing to be associated with partial charges of -0.35 or lower. Knowledge of the solid state will some day include an understanding of why each compound has its particular preference of crystal structure. The concept of partially charged atoms appears to offer more promise in this direction than does the ionic model.

In this connection, a rough relationship can be observed between the partial charge on a nonmetal atom and its coordination number in the crystal. For example, the partial charge on oxygen is about -0.8 in the alkali metal oxides where the coordination number of oxygen is 8 (except for Cs_2O). Where the partial charge on oxygen is around -0.6, in the M2 oxides, from MgO on, the oxygen coordination number is 6. Lower partial charges on oxygen correspond to oxides with 4:4 coordination as in ZnO, or rutile structures where the coordination number of oxygen is 3. The relationship is quite imperfect, because the natural combining capacity of the atoms also must be taken into account. However, the trend is consistent with the coordinated polymeric model in the sense that an atom coordinating

TABLE 9-4

ATOMIZATION ENERGIES OF SOLID M2 COMPOUNDS

Compound	Structure	R_o	n	M	k	t_i	E_c	E_i	Atomization energy (kpm)	
									Calc.	Exp.
BeF_2	β-Crystobalite	1.61	4	4.44	0.84	0.29	138.7	223.2	361.9	359.4
MgF_2	Dist. rutile	2.02[a]	4	4.76	0.84	0.41	89.0	269.4	358.4	352.0[b]
CaF_2	Fluorite	2.36	4	5.04	0.87	0.47	79.2	289.9	369.1	370.6
SrF_2	CaF_2	2.51	4	5.04	0.87	0.50	73.3	290.0	363.3	367.1
BaF_2	CaF_2	2.68	4	5.04	0.88	0.57	53.0	313.2	366.2	366.7
$BeCl_2$		2.02	3	4.09	0.85	0.23	130.8	131.4	262.2	258.9
$MgCl_2$	$CdCl_2$	2.54	3	4.49	0.87	0.34	83.1	173.6	256.7	256.9[b]
$CaCl_2$	Dist. rutile	2.74[c]	3.5	4.73	0.89	0.40	93.3	204.0	297.3	290.7
$SrCl_2$		3.02	4	4.62?	0.90	0.43	95.5	196.6	292.1	295.2
$BaCl_2$	$PbCl_2$	3.18	4	5.0	0.90	0.49	72.6	230.2	302.8	305.5
$BeBr_2$		2.16	3	4.09?	0.85	0.20	122.1	106.9	229.0	220.1
$MgBr_2$	CdI_2	2.70	3	4.38	0.88	0.31	77.3	146.9	224.2	222.9[b]
$CaBr_2$		2.94	3.5	4.73	0.89	0.36	87.1	171.1	258.2	257.3
$SrBr_2$		3.21[d]	4	4.62	0.90	0.39	90.0	167.7	257.7	263.6
$BaBr_2$	$PbCl_2$(?)	3.38	4	5.0	0.91	0.45	68.8	201.1	269.9	275.8
BeI_2		(2.37)	3	4.09(?)	0.87	0.14	115.8	69.8	185.6	179.9
MgI_2	CdI_2	2.94	3	4.38	0.89	0.24	74.4	105.6	180.0	180.0[b]
CaI_2	CdI_2	3.04	4	4.38	0.90	0.30	99.5	129.2	228.7	221.3
SrI_2		3.42	4	4.62(?)	0.91	0.33	87.1	134.7	221.8	225.5
BaI_2	$PbCl_2$	3.59	4	5.0	0.91	0.39	67.5	164.1	231.6	237.0
BeO		1.60	4	6.37	0.85	0.18	154.0	127.4	281.4	281.0
MgO	$NaCl$	2.10	3	1.75	0.82	0.25	102.9	142.9	245.8	248.9[b]
CaO	$NaCl$	2.40	4	1.75	0.85	0.28	100.7	145.2	245.9	253.9
SrO	$NaCl$	2.57	4	1.75	0.86	0.30	98.7	141.5	240.2	239.8
BaO	$NaCl$	2.76	4	1.75	0.86	0.34	75.2	155.1	230.3	235.0
BeS	Zinc blende	2.10	3	1.64	0.82	0.12	143.4	58.2	201.6	200.8
MgS	$NaCl$	2.60	3	1.75	0.86	0.19	98.8	83.3	182.1	192.6[b]
CaS	$NaCl$	2.84	4	1.75	0.87	0.22	132.5	89.3	221.8	219.1
SrS	$NaCl$	2.94	4	1.75	0.88	0.24	129.3	95.1	224.4	213.7
BaS	$NaCl$	3.18	4	1.75	0.88	0.28	101.2	96.5	197.7	214.6

[a] 1.98–2.05 Å.
[b] Based on assumption that atomization of energy of Mg is 45.5 instead of 35.5 kpm.
[c] 2.70–2.76 Å.
[d] 3.16–3.44 Å.

its outer shell electron pairs to its neighbors would be expected to be a better donor the higher its negative charge. One way of being a better donor would be to accommodate a larger number of acceptors.

SOLID COMPOUNDS OF M2′ ELEMENTS

Calculations of atomization energies of some of the compounds of zinc and cadmium are summarized in Table 9-5. Clearly something is wrong

TABLE 9-5

ATOMIZATION ENERGIES OF M2' SOLID COMPOUNDS

Com- pound	Struc- ture	R_o	n	M	k	t_i	E_c	E_i	$E_{calc.}$	$E_{exp.}$
ZnF_2		2.04	3	4.82	0.84	0.23	92.2	151.6	243.8	244.8
$ZnCl_2$	$CdCl_2$	2.64	3	4.49	0.87	0.16	109.7	78.6	188.3	188.6
$ZnBr_2$	$CdCl_2$	(2.75)	3	4.49	0.88	0.13	103.7	62.0	165.7	162.6
ZnI_2	CdI_2	2.95	3	4.38	0.89	0.07	99.6	30.7	130.3	132.0
ZnO		1.95	3	5.99	0.87	0.13	103.8	72.7	176.5	173.8
ZnS		2.33	3	6.55	0.87	0.07	137.2	32.4	169.6	145.8
If assumed nonpolar							147.7	0	147.7	145.8
CdF_2	CaF_2	2.32	3	5.04	0.87	0.27	71.9	169.4	241.3	229.8
										(245.5)
$CdCl_2$	$CdCl_2$	2.65	3	4.49	0.89	0.21	95.3	105.1	200.4	178.2
										(193.9)
$CdBr_2$	CdI_2	2.76	3	4.38	0.89	0.17	90.8	79.8	170.6	155.6
										(171.3)
CdI_2	CdI_2	2.98	3	4.38	0.90	0.12	84.6	52.7	137.3	126.0
										(141.7)
CdO	NaCl	2.35	3	1.75	0.89	0.16	78.0	88.8	166.8	147.5
										(163.2)
CdS		2.52	3	1.64	0.88	0.10	113.3	43.4	156.7	128.
If assumed nonpolar							126.0	0	126.0	128.

with most of the cadmium compound calculations. Somewhat better agreement could be obtained if the Cd single bond energy parameter were more like 25 instead of 35 kpm, but then cadmium would not fit on the $E = CrS$ line from bromine through tin. Alternatively, if the atomization energy of cadmium were 42.7 instead of 27, fair agreement would result, as shown in parentheses. Another puzzling observation is that no seemingly reasonable modification of calculations could produce atomization energies of the sulfides in even approximate agreement with the experimental values. Yet, as indicated in the table, if the compounds are treated as though they were completely nonpolar, the calculated atomization energies are in good agreement with the experimental values. Since the ionicity (t_i) is 7% for ZnS and 10% for CdS, it seems unreasonable to ignore it.

SOLID COMPOUNDS OF MANGANESE (II) AND LEAD (II)

In Table 9-6 are summarized the calculations of the atomization energies of some manganous and plumbous compounds. These are of course the results of a circular procedure, the experimental values having been used as a source of the metal bond energy parameters. They do demonstrate

TABLE 9-6

ATOMIZATION ENERGIES OF SOLID COMPOUNDS OF Mn(II) AND Pb(II)

Compound	Structure	R_o	n	M	k	t_i	E_c	E_i	$E_{calc.}$	E
PbF$_2$	PbCl$_2$	(2.29)	4	4.62	0.86	0.29	75.2	167.0	242.2	244.2
PbCl$_2$	PbCl$_2$	2.98^a	4	4.62	0.88	0.23	87.7	104.2	191.9	191.2
PnBr$_2$	PbCl$_2$	(3.19)	4	4.62	0.89	0.20	80.6	85.6	166.2	166.3
PbI$_2$	CdI$_2$	3.13	4	4.38	0.90	0.14	83.7	58.6	142.3	139.7
PbS	NaCl	2.97	4	1.75	0.88	0.11	101.1	43.2	144.3	135.9
MnF$_2$		2.12^b	4	4.82	0.85	0.33	84.6	211.7	296.3	294.8
MnCl$_2$		2.58	4	4.49	0.88	0.27	107.8	137.3	245.1	240.4
MnBr$_2$	CdI$_2$	2.70	4	4.38	0.89	0.23	103.3	110.2	213.5	209.4
MnI$_2$	CdI$_2$	2.96	4	4.38	0.89	0.17	97.6	74.3	171.9	175.0
MnO	NaCl	2.22	4	1.75	0.86	0.20	91.8	113.4	205.2	218.6
MnS	NaCl	2.61	4	1.75	0.86	0.14	125.3	61.2	186.5	182.6

^{a} 2.67–3.29 Å.
^{b} 2.10–2.13 Å.

however, that one could have chosen any one compound and then from its data consistently evaluate reasonably correctly the atomization energies of all the others of the same element.

SOLID HALIDES OF THALLIUM (I), COPPER (I), AND SILVER (I)

Table 9-7 summarizes the data for these compounds. Again they are the

TABLE 9-7

ATOMIZATION ENERGIES OF SOLID HALIDES OF Tl(I), Cu(I), AND Ag(I)

Compound	Structure	R_o	n	M	k	t_i	E_c	E_i	$E_{calc.}$	$E_{exp.}$
TiF	NaCl	2.59	4	1.75	0.88	0.49	44.8	96.7	141.5	140.1
TlCl	CsCl	3.31	4	1.76	0.91	0.40	57.7	64.3	122.0	121.7
TlBr	CsCl	3.44	4	1.76	0.91	0.36	55.8	55.6	111.4	111.5
TlI	CsCl	3.65	4	1.76	0.91	0.28	56.3	40.8	97.1	98.9
CuF		1.85	4	1.64	0.83	0.38	70.2	92.8	163.0	159.
CuCl		2.35	4	1.64	0.86	0.29	89.4	57.8	147.2	142.7
CuBr		2.46	4	1.64	0.87	0.25	85.4	48.1	133.5	132.8
CuI		2.62	4	1.64	0.88	0.17	84.6	31.1	115.7	122.7
AgF	NaCl	2.46	4	1.75	0.87	0.38	53.7	78.1	131.8	135.5
AgCl	NaCl	2.77	4	1.75	0.89	0.30	75.8	56.0	131.8	127.5
AgBr	NaCl	2.89	4	1.75	0.89	0.25	76.5	44.7	121.2	118.7
AgI		2.80	4	1.64	0.90	0.17	80.4	29.8	110.2	108.3

results of a circular procedure and demonstrate the internal consistency which is a minimum requirement for their tentative acceptance.

MISCELLANEOUS BINARY SOLIDS

The data summarized in Table 9-8 bring to over one hundred the number of binary solids for which the atomization energies are calculated and discussed in this chapter. Some of these data of Table 9-8 reveal an interesting new point. The rutile structure is one of 6:3 coordination. That is, the metal atom is octahedrally surrounded by six nonmental atoms. Each of the nonmetal atoms is coordinated to three metal atoms. Two alternative methods of dealing with such structures are possible. One is to treat them

TABLE 9-8

ATOMIZATION ENERGIES OF MISCELLANEOUS SOLIDS

Compound	Structure	R_o	n	M	k	t_i	E_c	E_i	$E_{calc.}$	$E_{exp.}$
$AlCl_3$		(2.16)	3	8.30	0.87	0.19	129.6	210.8	340.4	333.6
GeO_2	Rutile	1.89	6	2.41	0.85	0.06	230.9	108.8	339.7	340.9
Same, alternative calculation[a]							38.5	10.5	294.0	340.9
SnO_2	Rutile	2.05	6	2.41	0.85	0.09	210.0	150.5	351.5	330.0
Same, alternative calculation							33.5	14.6	288.6	330.0
PbO_2	Rutile	2.16	6	2.41	0.86	0.09	146.1	144.6	290.7	232.3
Same, alternative calculation							24.4	13.8	229.2	232.3
TeO_2	Rutile	2.20	6	2.41	0.85	0.06	183.9	95.3	279.2	241.8
Same, alternative calculation							30.6	9.2	238.8	241.8
TiO_2	Rutile	1.94	6	2.41	0.84	0.20	109.6	350.0	459.6	458.0
Same, alternative calculation							18.3	34.2	315.0	458.0
TiS_2	CdI_2	2.42	6	2.19	0.88	0.15	137.2	181.0	318.2	326.2

[a] The normal calculation is based on the coordinated polymeric model and uses the Born–Mayer expression for the ionic contribution. The alternative calculation simply assumes that each metal atom forms six polar covalent bonds to oxygen, which must be broken for atomization to occur.

as partially ionic substances for which the ionic part of the energy can be calculated by use of the Born-Mayer expression. The other is to treat them as covalent polymers with six single polar covalent bonds formed by each metal atom. The atomization energies calculated by the two methods are quite different. Both methods have been outlined in the table. It is interesting to observe that of the five rutile structured dioxides listed, TiO_2, GeO_2, and SnO_2 appear to fit the first above interpretation whereas PbO_2 and TeO_2 appear to fit the second. Except for SnO_2 which gives poor agreement either way, there seems to be little question as to which interpretation is appropriate. It is not easy to see how the two bonding interpretations can give different physical pictures. Here is one more indication of further work to be done.

ADVANTAGES OF THE COORDINATED POLYMERIC MODEL

The principal advantages of the coordinated polymeric model over the ionic model of binary nonmolecular solids may be stated as follows:

1. It provides a uniform concept of bonding applicable to nonmolecular compounds over a complete range of bond polarity.

2. It permits quantitative evaluation of the magnitudes of the separate covalent and ionic contributions to the total bonding energies.

3. With appropriate modification it permits calculation of bond energies in molecular compounds as well, offering a unified concept of bonding. The dichotomy of "ionic" and "covalent" bonds becomes largely unnecessary.

4. It gives a plausible explanation for the observed differences between experimental crystal radii and "ionic" radii, the actual crystal radii being those of only partially charged atoms.

5. Without being inconsistent with the known properties of nonmolecular solids, it permits a reasonable distinction to be made among different degrees of ionicity, such as have been so usefully correlated with the physical and chemical properties of the compounds.[8]

6. It takes quantitatively into account the inherent donor properties of anions and acceptor abilities of cations, thus presenting a more reasonable physical picture of the state of the atoms in the solid.

7. It suggests a reasonable fundamental basis for differences in crystal structure which cannot be accounted for by the ionic model.

8. It suggests that differences in solubility of nonmolecular solids may be related to the energy required to transform the polar covalent solid to ions, as well as to the ionic lattice energy and the solvation energy of the ions.

9. It lends credibility to the concept of partial charge, making the general use of this concept more acceptable as a basis for interpreting chemistry.

No doubt many competent chemists will still cringe at the thought of calcium oxide seriously being described as only 28% ionic, for example, or magnesium chloride being only 34% ionic. But the step from solid magnesium chloride to magnesium ions and chloride ions in solution is not nearly as drastic as it might first appear. The magnesium ions are merely changing from one kind of coordination sphere to another. The concept of ions in a crystal is not objectionable as long as one keeps in mind what ions would do in a crystal—namely, coordinate.

As suggested earlier in this chapter, there are two ways in which a given

[8] R. T. Sanderson, "Chemical Periodicity". Reinhold, New York, 1960; Ibid., "Inorganic Chemistry". Reinhold, New York, 1967.

cation might be attracted to an anion. One way would involve coordination as described, in which outermost electrons of the anion do penetrate toward the nucleus of the "cation," thus destroying its identity as a cation. The other way would involve some condition wherein penetration of anion electrons toward the nucleus of the cation would be unfeasible. There would still exist a substantial electrostatic attraction between the cation and the electron pairs of the anion. In other words, the coordinated polymeric model does not altogether rule out the ionic model except for simple binary compounds. For a compound such as tetramethylammonium chloride, for example, an ionic model appears to be quite reasonable and a coordinated polymeric model perhaps less so. Then there are also a number of compounds in which equalization of electronegativity would seem impossible. For example, the geometric mean electronegativity of sodium sulfate is higher than the electronegativity of sodium ion. Under such circumstances, it seems reasonable to suppose that the sodium atom loses its electron completely and the compound is ionic. However, compounds of this type have not yet been studied from the viewpoint of atomization energies.

Chapter 10

APPLICATIONS TO ORGANIC CHEMISTRY

C—C and C—H Bond Energies

Appreciable differences are observed among very carefully measured standard heats of formation of gaseous hydrocarbon isomers.[1] For example, the values for normal and isobutane are reported as -29.81 and -31.45 kpm. Accurate data are available for 75 decane isomers. Their heats of formation range from -58.57 to -68.27 kpm. This presents a problem in the application of the bond energy calculation method to organic chemistry, for the method makes no allowance for any differences among isomers. Only the empirical formula is needed for calculating the geometric mean electronegativity and the partial charges on the individual atoms. No distinction is made between individual C—H or C—C bonds unless different bond lengths are reported.

Small differences in C—C single bond lengths have occasionally been observed, but usually the measurements are not sufficiently accurate to judge the significance of the differences. Similarly, differences in C—H bond lengths have been noted. Very careful studies of propane and isobutane[2] revealed the following differences in C—H bond length (in Å): 1° hydrogen, 1.091; 2° hydrogen, 1.096; and 3° hydrogen, 1.108. These imply a small progressive weakening of the bond in that order, as will be considered presently.

In general, calculation of the atomization energies of organic molecules presents a more difficult problem than provided by the average inorganic molecule because of the common occurrence of relatively large numbers of like bonds. One might be moderately content with an error per bond energy of 1 or 2 kpm provided only a few such bonds occur in the molecule. But

[1] Comprehensive Index of API 44-TRC Selected Data on Thermodynamics and Spectroscopy, Thermodynamics Research Center, Dept. of Chemistry, Texas A & M University, College Station, Texas, 1968.

[2] D. R. Lide, *J. Chem. Phys.* **33,** 1514, 1519 (1960).

even a simple butyl compound has nine C—H bonds, and any error in the bond energy calculation is quickly multiplied to a total value large enough to place a severe limit on the utility of the method. The differences among C—H and C—C bond energies may seem negligibly small from the viewpoint of the overall atomization energy. For example, the 1.64 kpm difference between normal butane and isobutane seems quite tolerable when divided up among the thirteen bonds per molecule, giving an average difference of only 0.13 kpm per bond. A comparison of the total atomization energies of 1235.9 and 1237.6 kpm looks almost like perfect agreement. Even in the decanes, which contain thirty-one bonds per molecule, the average energy difference per bond is 0.31 kpm. But when translated to a difference of nearly 10 kpm in the standard heats of formation the situation looks much less attractive. At least the problem needs evaluating to determine where the greatest difficulties lie and how serious they are.

Much detailed attention has been devoted to this general area, of which the references are representative.[3] Previous studies have been both empirical and theoretical. Without the need at present of a thorough discussion of this work, we can gain some useful ideas by examining a few of the available data.

It would perhaps be unnecessarily laborious to study in detail all the data for the 75 decanes. The general principles can be recognized from comparison of the structures and heats of formation of the gas phase nonanes. Figure 10-1 presents carbon skeletons of 35 nonane isomers in order of decreasing standard heat of formation. Two factors may be pointed out. One has already been mentioned in discussion of the choice of a carbon single bond energy parameter. The single bond energy in diamond was shown to be 85.4 kpm yet, on empirical grounds, the parameter chosen was 83.2 kpm. In some manner the C—C bond energy seems to be enhanced by structures in which all four bonds formed by a given carbon atom join to other carbon atoms. The bonds in hydrocarbons appear to be stronger, the more nearly the condition of the carbon atom resembles that in diamond. Strength would then increase in the order 1° carbon to 2° carbon to 3° carbon to 4° carbon.

[3] C. T. Zahn, *J. Chem. Phys.* **2**, 671 (1934); J. R. Platt, *J. Chem. Phys.* **15**, 419 (1947); J. R. Platt, *J. Phys. Chem.* **56**, 328 (1952); R. D. Brown, *J. Chem. Soc.* p. 2615 (1953); M. J. S. Dewar and R. Pettit, *J. Chem. Soc.* p. 1625 (1954); K. J. Laidler, *Can. J. Chem.* **34**, 626 (1956); J. B. Greenshields and F. D. Rossini, *J. Phys. Chem.* **62**, 271 (1958); T. L. Allen, *J. Chem. Phys.* **31**, 1039 (1959); E. G. Lovering and K. J. Laidler, *Can. J. Chem.* **38**, 2367 (1960); V. M. Tatevskii, V. A. Benderskii, and S. S. Yarovoi, "Rules and Methods for Calculating the Physico-Chemical Properties of Paraffinic Hydrocarbons" (translation ed.) (B. P. Mullins, ed.). Pergamon Press, Oxford, 1961; H. A. Skinner and G. Pilcher, *Quart. Rev.* **17**, 264 (1963). Also J. W. Anderson, G. H. Beyer, and K. M. Watson, *Nat. Petrol. News* **36**, R 476 (1944); J. L. Franklin, *Ind. Eng. Chem.* **41**, 1070 (1949); M. Souders, C. S. Mathews, and C. O. Hurd, *Ind. Eng. Chem.* **41**, 1408 (1949); S. W. Benson and J. H. Buss, *J. Chem. Phys.* **29**, 546 (1958).

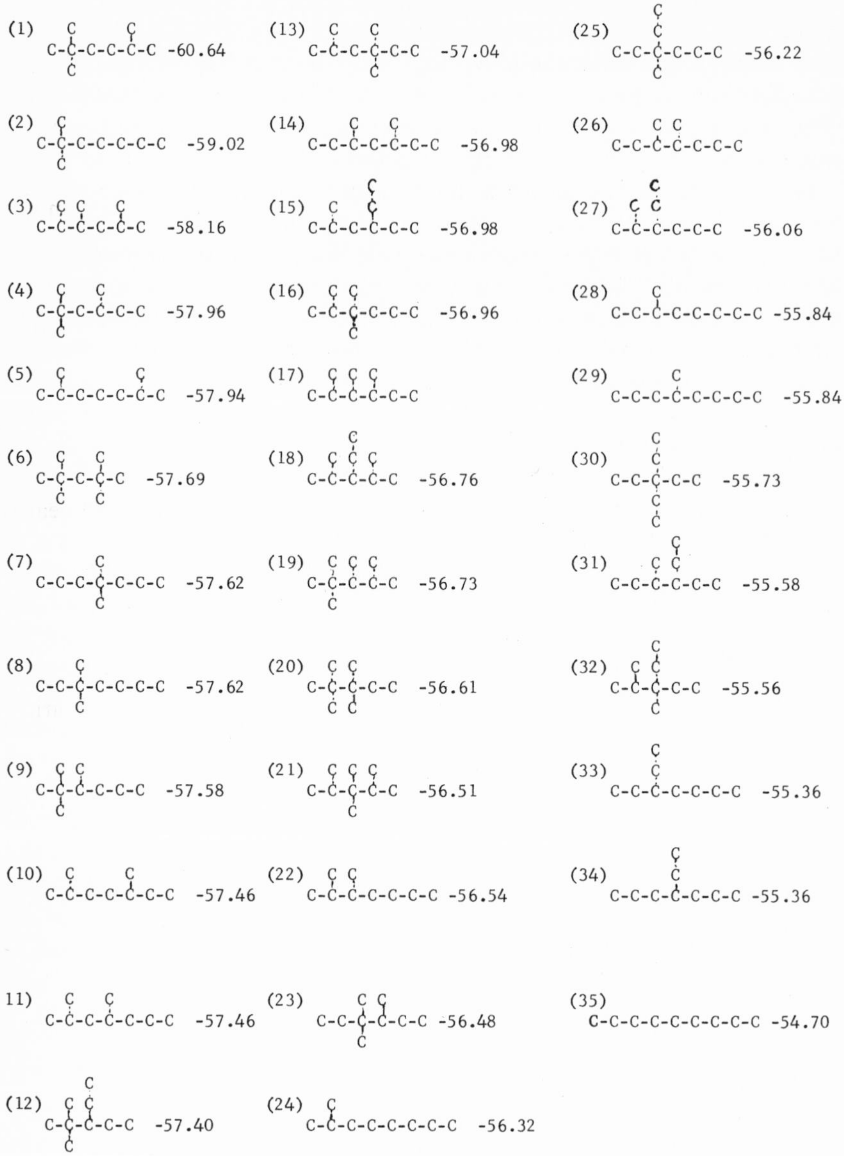

(1)   C   C
     C-C-C-C-C-C -60.64
         C

(2)   C
     C-C-C-C-C-C -59.02
       C

(3)   C C   C
     C-C-C-C-C -58.16

(4)   C   C
     C-C-C-C-C -57.96
       C

(5)   C   C
     C-C-C-C-C-C -57.94

(6)   C   C
     C-C-C-C -57.69
       C   C

(7)   C
     C-C-C-C-C-C -57.62
           C

(8)   C
     C-C-C-C-C-C -57.62
         C

(9)   C C
     C-C-C-C-C -57.58
       C

(10) C   C
     C-C-C-C-C-C -57.46

11)   C   C
     C-C-C-C-C-C -57.46

(12)   C
      C C
     C-C-C-C -57.40
       C

(13) C   C
     C-C-C-C-C -57.04
         C

(14)   C   C
     C-C-C-C-C-C -56.98

(15)   C
      C   C
     C-C-C-C-C -56.98

(16) C C
     C-C-C-C-C -56.96
         C

(17)   C C C
     C-C-C-C-C-C

(18)   C
      C C C
     C-C-C-C -56.76

(19)   C C C
     C-C-C-C -56.73
       C

(20)   C C
     C-C-C-C -56.61
       C C

(21)   C C C
     C-C-C-C -56.51
         C

(22)   C C
     C-C-C-C-C-C -56.54

(23)   C C
     C-C-C-C-C-C -56.48
         C

(24)   C
     C-C-C-C-C-C-C -56.32

(25)   C
           C
     C-C-C-C-C-C -56.22
         C

(26)   C C
     C-C-C-C-C-C

(27)   C
      C C
     C-C-C-C-C-C -56.06

(28)   C
     C-C-C-C-C-C-C -55.84

(29)   C
     C-C-C-C-C-C-C -55.84

(30)   C
           C
     C-C-C-C-C -55.73
           C
           C

(31)   C
      C C
     C-C-C-C-C-C -55.58

(32)   C
      C C
     C-C-C-C-C -55.56
         C

(33)   C
           C
     C-C-C-C-C-C-C -55.36

(34)   C
            C
     C-C-C-C-C-C-C -55.36

(35)
     C-C-C-C-C-C-C-C-C -54.70

FIG. 10-1. Structures and heats of formation of gaseous nonane isomers.

The other factor appears to be steric, resulting from the attached hydrogen atoms. It weakens, at least in effect, the carbon–carbon bonds. The decanes give a more spectacular example than the nonanes of the way in which these two factors operate, sometimes in opposition. The most stable of the

75 decanes reported is 2,2,5,5-tetramethylhexane [$\Delta Hf°(g)$, -68.27]. The least stable is not n-decane [$\Delta Hf°(g)$, -59.63] as might be expected, but 2,2,3,3,4-pentamethylpentane [$\Delta Hf°(g)$, -58.57]. The first structure has two $(CH_3)_3C$— groups separated by two carbon atoms where they do not interfere with one another. The second structure has two quaternary carbon atoms adjacent where their substituents interfere. In this particular compound, the weakening ascribable to steric effects appears more than sufficient to overbalance the strengthening associated with the quaternary carbon atoms. More commonly, the weakening is not that great. A study of space-filling molecular models reveals very clearly the steric differences between these structures, identifying those structures where reduction of the heat of formation would be expected.

In the nonanes of Fig. 10-1, the most stable isomer (1) is that which comes closest to having two $(CH_3)_3C$— groups separated by two carbon atoms as in the most stable decane. Isomer (6) shows that a one carbon atom separation is not enough. This is consistent with the shifting of the lone methyl group from the 5 position in isomer (1) to the 4 position in isomer (4) and the 3 position in isomer (9). Numerous other related trends and tendencies are also revealed by Fig. 10-1, but there is no need to discuss them all. Primarily one may expect carbon–carbon branching to increase the total bonding energy somewhat, except that such an effect would be reduced and perhaps even eliminated, in effect, by steric interference of the branches.

ATOMIZATION OF METHANE

In methane the partial charges are approximately 0.01 on hydrogen and -0.05 on carbon. In general, in hydrocarbons and their derivatives, the carbon charge is very close to 0.06 more negative than the hydrogen charge, even when both are positive, giving a constant value of 0.03 for t_i. The hydrogen bond energy parameter is corrected for the positive charge on hydrogen by multiplying by 0.99, giving 103.2. Taking 83.2 as the carbon bond energy parameter, one calculates the geometric mean energy as 92.8 kpm. The observed bond length is the same as the covalent radius sum, 1.09 Å. The covalent energy contribution is then

$$t_c E_c = \frac{0.97 \times 92.8 \times 1.09}{1.09} = 90.0 \text{ kpm}$$

The ionic contribution is

$$t_i E_i = \frac{0.03 \times 332}{1.09} = 9.1 \text{ kpm}$$

Their sum, 99.1, is the calculated C—H bond energy in CH_4. Multiplied by 4, this is 396.4 kpm, to be compared with the experimental value of

397.6 kpm. Even though the agreement is within 0.3%, the value is evidently a little low, by about 0.3 kpm of C—H bonds. The calculation demonstrates, however, that the relative electronegativities, partial charges, bond polarity, and covalent and ionic contributions must be reasonably accurate. In other words, carbon and hydrogen present no special problem here.

ATOMIZATION OF ALKANES AND THEIR ISOMERS

The charge distribution between carbon and hydrogen changes from methane to the higher alkanes as the formula approaches CH_2. For this empirical formula, the partial charge on hydrogen is 0.02 and on carbon, -0.04. The corrected hydrogen bond energy parameter is then about 102 kpm, giving a geometric mean of 92.2 with the carbon parameter of 83.2. The value of 83.2 is considered to be the "normal" value, with small increases the result of attachment to like atoms in a manner providing greater electron correlation. Using the bond lengths for primary, secondary, and tertiary hydrogen as given by Lide,[2] one can then calculate the following bond energies: C—H′, 98.5; C—H″, 98.2; and C—H‴, 97.1 kpm.

TABLE 10-1
EMPIRICAL BOND ENERGIES APPLICABLE TO ALKANES

Bond type	Energy (kpm)
C—H′	98.5
C—H″	98.2
C—H‴	97.1
C′—C′	84.4
C′—C″	84.1
C′—C‴	84.7
C′—C⁗	84.8
C″—C″	84.1
C″—C‴	84.4
C″—C⁗	84.8
C‴—C‴	84.6

These C—H bond energies may then be used to calculate C—C energies from the experimental atomization energies of the different alkane isomers. For example, the C′—C⁗ energy of 84.8 kpm is calculated from neopentane, by dividing by 4 the difference between the experimental atomization energy of 1521.3 and $12 \times 98.5 = 1182.0$. The C′—C″ energy similarly calculated from propane is 84.1, which is then found to be the same for C″—C″ in normal butane. These results are summarized in Table 10-1. Included in Table 10-2 are the experimental atomization energies of a number of branched and straight chained alkanes, together with the values calculated as the sum of the individual bond energies. The system is obviously imperfect,

TABLE 10-2

ATOMIZATION ENERGIES OF SOME ALKANES AND THEIR ISOMERS

Compound	Atomization energy (kpm)	
	Calc.	Exp.
Ethane	675.4	675.4
Propane	955.5	955.5
n-Buthane	1236.1	1236.1
Isobutane	1237.6	1237.6
n-Pentane	1516.6	1516.6
Isopentane	1517.9	1518.5
Neopentane	1521.3	1521.3
n-Hexane	1797.1	1797.0
2-Methylpentane	1798.4	1798.7
2,2-Dimethylbutane	1801.5	1801.4
3-Methylpentane	1798.1	1798.1
2,3-Dimethylbutane	1799.6	1799.6
n-Heptane	2077.6	2077.5
2-Methylhexane	2078.9	2079.2
3-Methylhexane	2078.6	2078.5
3-Ethylpentane	2078.3	2077.9
2,2-Dimethylpentane	2081.5	2081.9
2,3-Dimethylpentane	2079.8	2080.2
2,4-Dimethylpentane	2080.2	2080.9
3,3-Dimethylpentane	2081.2	2080.7
2,2,3-Trimethylbutane	2083.2	2081.5
2,2-Dimethylheptane	2642.5	2642.5
n-Decane	2919.1	2918.7
n-Pentadecane	4321.6	4320.7
n-Eicosane ($C_{20}H_{42}$)	5724.1	5722.7

but it does provide a rough and ready method of estimating the atomization energies or heats of formation of these gaseous compounds with quite satisfactory accuracy. The experimental values will always be lower than the calculated values, however, whenever opportunity for steric interference exists. Such interference appears to begin when a methyl group branch is attached to a carbon that is adjacent to a nonterminal carbon that already has two methyl branches. In Table 10-2 this effect is shown in triptane, 2,2,3-trimethylbutane, to provide an example. Other such isomers were deliberately omitted. They are numerous and no attempt is made here to evaluate quantitatively the steric effects.

C—H AND C—C BOND ENERGIES IN HYDROCARBON DERIVATIVES

Most hydrocarbon derivatives, indeed practically all except organo-metallic compounds, contain functional groups that are initially more electronegative than carbon and hydrogen, and therefore withdraw electrons from them. The partial negative charge on carbon is so small that only a slight withdrawal of electrons will give it a partial positive charge. In most organic derivatives, therefore, we are faced with the problem of how to treat the polarity of a bond between two atoms, both bearing partial positive charge. It has proved satisfactory when the two positively charged atoms are alike to reduce their normal nonpolar covalent energy by the factor $(1.00 - \delta_E)$. This takes into account the bond weakening expected relative to a similar bond between two neutral atoms. For the carbon–hydrogen bond, the hydrogen parameter is corrected in the customary manner by the factor $(1.00 - \delta_H)$. The equation for t_i appears applicable even when both atoms bear the same sign of partial charge, for t_i is half the *difference* between the two charges. Thus if the charge on carbon is $+0.03$ and that on hydrogen -0.03, $t_i = [0.03 - (-0.03)]/2 = 0.03$. If the charge on carbon is 0.03 and that on hydrogen 0.09, $t_i = (0.09 - 0.03)/2 = 0.03$.

Other kinds of bonds in organic molecules in general give no difficulty in bond energy calculation. One can calculate the atomization energies of most organic compounds with reasonable accuracy by merely assigning the values of 99 to the C—H bond and 83 to the C—C bond and calculating the energies of the other bonds in the usual manner. An empirical attempt to refine this procedure somewhat consisted of using the experimental atomization energies of a large number (about 60) of representative organic compounds to back-calculate by difference the C—H bond energies. This procedure resulted in the adoption of the following average values for C—H energy: 98.7 except for tertiary hydrogen and hydrogen attached to aromatic carbon; 97.3 for hydrogen attached to aromatic carbon; tertiary hydrogen to be calculated separately as the usual sum of a covalent and an ionic contribution. These values are used for the calculations to be discussed in the following sections.

It should be noted that no account is taken of the possible existence of molecular stabilization through internal protonic bridging, which in some compounds might add appreciably to the experimental atomization energy. It must also be kept in mind that the exact constancy of C—C and C—H bond energies cannot be expected since these values do vary somewhat. However, the method is nevertheless sufficiently accurate to provide a reasonably quantitative explanation of bond energies and heats of formation with the many applications these data have.

Atomization Energies of Some Organic Compounds

Since there is no essential difference between organic molecules and any other kind of molecule with respect to bond energy calculations, perhaps one detailed example can be discussed to illustrate this, and various other results can be summarized in table form (see Table 10-3).

Acetic acid, CH_3COOH, has the empirical formula $C_2H_4O_2$. From the atomic electronegativities one can calculate the geometric mean to be 3.97 for the molecule. This corresponds to partial charges of 0.11 on hydrogen, 0.05 on carbon, and -0.26 on oxygen. Five kinds of chemical bond are incorporated into this one molecule, for which the energy may be calculated separately.

The value of 98.7 for C—H energy is adopted here. Three such bonds contribute 296.1 kpm to the total atomization energy.

The carbon–carbon bond length is reported to be 1.50 Å instead of the covalent radius sum of 1.54 Å. Since each carbon bears a partial charge of 0.05, the single bond energy must be reduced by the factor 0.95. The carbon–carbon energy is then

$$E = \frac{0.95 \times 83.2 \times 1.54}{1.50} = 81.1 \text{ kpm}$$

For the O—H bond, t_i is 0.18, and the bond length 0.95 Å. The corrected hydrogen bond energy parameter is 92.8, which with the oxygen energy of 34.0 kpm gives a geometric mean of 56.3 kpm. The covalent radius sum is 1.04 Å. The covalent energy contribution is

$$t_cE_c = \frac{0.82 \times 56.3 \times 1.04}{0.95} = 50.5 \text{ kpm}$$

The ionic energy is

$$t_iE_i = \frac{0.18 \times 332}{0.95} = 62.9 \text{ kpm}$$

The sum, 113.4 kpm, is the O—H bond energy in acetic acid.

There are two carbon–oxygen bonds per molecule. The reported length of the C—OH bond is 1.31 Å, to be compared with 1.42 Å for the carbon–oxygen bond in dimethylether and 1.49 for the covalent radius sum. The reported length of the C=O bond in acetic acid is 1.25 Å, to be compared with the C=O bond length of 1.22 Å in both acetaldehyde and acetone, and the covalent radius sum of 1.49 Å. These evidences of "resonance" suggest that the oxygen parameter should be the double prime value, O″ 66.7, for *both* carbon–oxygen bonds. The nonpolar covalent C—O″ energy is then 74.5. For the single bond, the covalent energy is

$$t_cE_c = \frac{0.85 \times 74.5 \times 1.49}{1.31} = 72.0 \text{ kpm}$$

The ionic energy is

$$t_iE_i = \frac{0.15 \times 332}{1.31} = 38.0 \text{ kpm}$$

The total bond energy is the sum 110.0 kpm.

For the double bond, the multiplicity factor of 1.50 is used, and the covalent energy contribution is:

$$t_c E_c = \frac{0.85 \times 74.5 \times 1.50 \times 1.49}{1.25} = 113.2 \text{ kpm}$$

The ionic energy is

$$t_i E_i = \frac{0.15 \times 332 \times 1.50}{1.25} = 59.8 \text{ kpm}$$

The sum 173.0 is the energy of the $C{=}O$ bond.

The total atomization energy is the sum of these calculated bond energies, which is 773.6 kpm. The experimental value is 774.9 kpm. These correspond to a calculated and experimental heat of formation of -103.4 and -104.7 kpm.

In a similar way, the results given in Table 10-3 were obtained. Full details are given in the comprehensive table in the appendix (see Table B, p. 200). Calculations have been completed for most of the organic hydrocarbon derivatives for which gas phase heats of formation are known as well as for many others too unstable to exist in the gas state. In general the results are excellent, and bring to organic chemistry the same kind of insights provided to inorganic chemistry.

For example, the advantages of double bonding to oxygen over single bonding reflected in the instability of carbonic acid show also why in general two hydroxyl groups on the same carbon atom tend to split out water and form a carbonyl group. The same basis for explaining the monomeric nature of CO_2 as opposed to the polymeric nature of SiO_2 serves also to explain the nonexistence of acetone polymer analogous to the dimethylsilicones. By the use of these methods, cohesive forces in liquids and solids incapable of existing in the vapor state can be estimated. One can calculate the atomization energy of the enol form of acetone, and by comparing it with that of the keto form discover the latter to be more stable by about 20 kpm.

Unquestionably these procedures of bond energy calculation are rich with possibilities of enhancing the understanding of organic chemistry, and deserve much further study.

Standard Bond Energies

Empirically derived standard bond energies which are additive throughout a molecule, and therefore capable of furnishing reasonable estimates of atomization energies and heats of formation, have long been available. Their major application seems to have been to organic compounds. They

TABLE 10-3

CALCULATED AND EXPERIMENTAL ATOMIZATION ENERGIES
AND STANDARD HEATS OF FORMATION OF SOME REPRESENTATIVE ORGANIC MOLECULES

Compound	Formula	Atomization energy (kpm)		Heat of formation	
		Calc.	Exp.	Calc.	Exp.
Methyl chloride	CH_3Cl	379.1	376.3	-22.4	-19.6
Methyl mercaptan	CH_3SH	455.1	449.3	-8.8	-3.0
Methyl amine	CH_3NH_2	557.9	551.5	-13.1	-6.7
Ketene	$CH_2{=}C{=}O$	523.0	521.0	-16.6	-14.6
Acetonitrile	CH_3CN	588.7	590.9	$+23.2$	$+21.0$
Acetaldehyde	CH_3CHO	654.1	650.4	-43.5	-39.8
Ethylbromide	C_2H_5Br	646.1	642.8	-16.3	-13.0
Ethyl alcohol	C_2H_5OH	771.3	770.8	-56.5	-56.0
Dimethylether	CH_3OCH_3	755.8	759.1	-41.0	-44.3
Ethylene glycol	$CH_2(OH)CH_2(OH)$	862.5	866.9	-88.1	-92.5
Dimethylamine	$(CH_3)_2NH$	827.5	826.9	-7.2	-6.6
Acetone	$(CH_3)_2CO$	933.8	938.5	-47.7	-52.4
Propionaldehyde	C_2H_5CHO	936.5	935.3	-50.4	-49.2
Propionic acid	C_2H_5COOH	1056.0	1054.5	-110.3	-108.8
Allylamine	$CH_2{=}CHCH_2NH_2$	981.8	984.8	$+9.8$	$+6.8$
Ethylacetylene	$HC{\equiv}CC_2H_5$	958.3	959.8	$+39.5$	$-38.$
1,3-Butadiene	$CH_2{=}CH{-}CH{=}CH_2$	967.2	970.3	$+30.6$	$+27.5$
Methylacrylate	$CH_2{=}CHCOOCH_3$	1181.2	1187.1	-64.2	-70.1
Acetic anhydride	$(CH_3CO)_2O$	1329.0	1325.4	-152.4	-148.8
Ethyl propionate	$C_2H_5COOC_2H_5$	1602.5	1609.1	-105.8	-112.4
Benzene	C_6H_6	1319.4	1320.6	$+21.0$	$+19.8$
Phenol	C_6H_5OH	1421.1	1421.7	-21.1	-21.7
Cyclohexane	C_6H_{12}	1683.6	1682.4	-30.6	-29.4
Hexamethylenediamine	$(CH_2)_6(NH_2)_2$	2123.3	2118.0	-35.9	-30.6
Benzaldehyde	C_6H_5CHO	1581.4	1580.9	-10.1	-9.6
Toluene	$C_6H_5CH_3$	1603.1	1603.7	$+12.8$	$+12.2$
Styrene	$C_6H_5CH{=}CH_2$	1748.1	1752.6	$+39.1$	$+34.6$
Cyclooctatetraene	C_8H_8	1718.8	1717.1	$+68.3$	$+70.$

have been recognized as having only approximate validity but in general some practical utility. The bond energy calculation methods described in this book afford the first possibility of a fundamental explanation of both the quantitative values and the approximate additivity of these standard energies.

First, the concept that useful "standard" energies could exist seems in contradiction to the belief that each bond in a molecule must be affected by all the other bonds. What basis is there, for example, for assuming that the O—H bonds in hydrogen peroxide will have exactly the same energy as those in water. The answer seems to lie in the nearly constant values of

t_i, the ionicity coefficient. The nature of electronegativity equalization is such that the partial charges on any two different atoms in a molecule tend to move upward or downward at a nearly equal rate in different compounds, such that their average remains almost constant. For example, a rather extreme one, the partial charges on carbon and fluorine in the CF molecule are 0.22 and -0.22, giving $t_i = 0.22$. In CF_2 the charges on carbon and fluorine are 0.30 and -0.15, from which $t_i = 0.23$. In CF_3 the charges are 0.34 and -0.11, and in CF_4, 0.37 and -0.09, still leading to $t_i = 0.23$. This is the reason why, although each bond in a molecule does indeed affect each other bond, the energy of a bond between two given elements tends to remain approximately the same in a variety of environments. This is why "standard" bond energies have any validity at all.

TABLE 10-4

STANDARD AND CALCULATED CARBON–OXYGEN BOND ENERGIES

Bond type: C—O′		Bond type: C=O″	
Standard	84.0	Standard	
Calculated		HCHO	164
H_2CO_3	86.7	Other RCHO	171
$(CO_2)_x$	84.3	Ketones	174
CH_3OCH_3	81.8	Calculated	
CH_3OH	81.2	CO_2	192.1
C_2H_5OH	81.2	COS	189.3
$C_2H_5OC_2H_5$	81.2	$CH_2=C=O$	184.8
C_4H_9OH	80.7	C_3O_2	184.5
$HCOOCH_3$	80.7	$HCOOCH_3$	180.2
$(CH_2OH)_2$	81.2	HCHO	178.8
C_3H_7OH	80.1	$CH_2=CHCOOCH_3$	178.8
$CH_3OC_2H_5$	80.1	CH_3CHO	177.2
$CH_2=CHCOOCH_3$	79.6	$CH_2=CHCHO$	177.2
$CH_3COOC_2H_5$	79.6	C_2H_5CHO	177.2
$C_2H_5COOC_2H_5$	79.6	C_3H_7CHO	177.2
C_2H_5ONO	79.0	C_6H_5CHO	177.2
		HCOOH	175.8
Bond type: C—O″		$(CH_2)_4(COOH)_2$	175.7
Calculated			
HCOOH	111.7	H_2CO_3	175.6
$(CH_2)_4(COOH)_2$	111.7	C_2H_5COOH	173.1
CH_3COOH	110.0	$HCONH_2$	173.0
C_2H_5COOH	110.0	CH_3COOH	173.0
$(CH_3CO)_2O$	110.0	CH_3COCH_3	173.0
$CH_3COOC_2H_5$	110.0	$(CH_3CO)_2O$	173.0
$C_2H_5COOC_2H_5$	110.0	$CH_3COOC_2H_5$	173.0
$HCOOCH_3$	108.3	$C_2H_5COOC_2H_5$	173.0
$CH_2=CHCOOCH_3$	106.9		

As to the constancy of bond energies, Table 10-4 illustrates the situation with respect to the carbon–oxygen bond. Table 10-5 lists some other examples of how bond energy may vary from compound to compound. Finally, Table 10-6 permits a comparison among standard bond energies obtained from widely accepted literature sources[4] and by various methods of calculation. One of these methods of calculation is simply to assume the existence of a diatomic molecule, singly bonded. An average bond length

TABLE 10-5

MISCELLANEOUS CALCULATED BOND ENERGIES

Bond type: N—H		Bond type: C—N	
Standard	93.4	Standard	69.7
Calculated		Calculated	
NH	89.8	CH_3NH_2	72.2
NH_2	91.9	$(CH_3)_2NH$	72.6
NH_3	93.4	$C_2H_5NH_2$	72.2
N_2H_4	92.9	$(CH_3)_3N$	72.2
$HCONH_2$	91.6	$(CH_2)_6(NH_2)_2$	70.2
CH_3NH_2	94.8		
$(CH_3)_2NH$	90.1	Bond type: C=C	
$C_2H_5NH_2$	90.1	Standard	147
CH_2=$CHCH_2NH_2$	93.7	Calculated	
		C_8H_8	144.5
Bond type: O—H		CH_2=C=O	140.8
Standard	110.6	C_6H_5CH=CH_2	144.5
Calculated		CH_2=$CHCl$	142.0
HCOOH	114.8	1-Hexene	143.4
CH_3OH	113.4	CH_2=$CHCHO$	139.9
CH_3COOH	113.4	1-Pentene	143.4
C_2H_5OH	113.9	Propene	142.4
$(CH_2OH)_2$	111.6	Cyclopentadiene	142.4
C_2H_5COOH	114.2	1,3-Butadiene	143.4
C_3H_7OH	113.9		
C_4H_9OH	114.8	Bond type: C—Cl	
H_2CO_3	113.7	Standard	78.5
C_6H_5OH	113.6	Calculated	
$(CH_2)_4(COOH)_2$	113.9	$COCl_2$	86.8
HNO_3	112.7	CCl_4	86.9
H_2SO_4	112.8	$CHCl_3$	85.1
H_2O	111.8	CH_2Cl_2	84.6
H_2O_2	111.5	CH_3Cl	83.0
		CH_2=$CHCl$	87.0

[4] T. L. Cottrell, "Strengths of Chemical Bonds," 2nd Ed. Butterworths, London, 1958; L. Pauling, "Nature of the Chemical Bond," 3rd Ed. Cornell University Press, Ithaca, New York, 1960.

TABLE 10-6

STANDARD BOND ENERGIES—EMPIRICAL AND CALCULATED

Bond	Standard energy (kpm)				Average contributions	
	Lit.[a]	Exp.[b]	Calc.[c]	Calc.[d]	E_c	E_i
C—H	98.8	99.4	98.1	98.5	89.4	9.1
N—H	93.4	93.4	92.6	93·4	57.6	35.8
P—H	76.4	78.3	76.3	76.3	71.6	4.7
O—H	110.6	110.8	112.7	111.8	49.6	62.2
S—H	81–83	87.9	87.1	87.8	70.3	17.5
C—C	82.6	85.2	83.2	83.2	83.2	0
C=C	145.8	144	143.5	143.5	143.5	0
C≡C	199.6	—	189.0	189.0	189.0	0
C—O′	84.0	—	83.8	81	46	35
C=O″	192.0	192.3	192.1	192.1	119	73
C—S	62–65	—	71.9	71	64	7
C—Cl	78–81	78.1	84.1	85.2	61	24
C—Br	66–68	—	71.3	71	57	14
C—I	51–57	—	54.8	54.9	53	2
Si—F‴′	129.3	142.6	132.3	137.9	58	80
Si—Cl‴	85.7	95.6	94.0	95.0	51	44
N—F	64–65	66.7	67.4	68.9	35	34
P—Cl″	79.1	78.6	79.4	79.0	50	29
P—Br′	65.4	64.4	62.7	63.6	42.5	21
P—I′	51.4	46.9	47.0	46.9	40	7
O—F	44–45	44.9	47.7	48.0	34	14
O—Cl	48–52	49.3	49.0	49.3	43	6
S—Cl	59.7	64.8	66.2	67.2	52	15

[a] Text footnote 3.   [b] Experimental average atomization energy per bond.
[c] Calculated for hypothetical diatomic molecule.
[d] Average value calculated from actual compounds.

taken from the compilation of selected bond lengths[5] is assigned. Then the bond energy is calculated in the usual manner. Because of the approximate constancy of t_i as mentioned earlier, the values obtained in this way are in reasonably good agreement with those obtained by dividing the experimental atomization energy of a binary molecule by the number of bonds. But even though standard bond energies may have approximate validity for a considerable number of compounds, it would now seem quite unnecessary to rely upon them for making quick and useful estimates.

Table 10-6 also includes the approximate magnitudes of the covalent and ionic contributions to the bond energies, which is certainly useful information if the general nature of the bond is to be understood.

[5] *Interatomic Distances Supplement*, Special Publication No. 18, The Chemical Society, London, 1965.

Cohesive Energy

The bond energy methods described in this book are applicable directly only to the gaseous state of molecular compounds, and to nonmolecular solids. Their utility can be extended where reliable experimental cohesive energies for the molecular condensed phases are available, or where reasonable estimates of such cohesive energies can be made. Two excellent sources of thermochemical information have appeared recently.[6] Both give far more extensive compilations of heats of formation of organic compounds in both gaseous and condensed states than were available for the work described here. In addition, the book by Cox and Pilcher points out that energy differences between gaseous and condensed states at 25° are known experimentally for only about half the organic compounds for which thermochemical measurements have been made. They therefore review many previously published methods of estimation, several of which could be satisfactorily applied to obtain standard heats of formation of the liquid compounds from calculated gas-phase atomization energies. Of these, the simplest are the equations of Wadso,[7] who estimates ΔH_v at 25° as a simple linear function of the boiling point in °C. For slightly or nonassociated liquids he gives:

$$\Delta H_v = 5.0 + 0.041 t_b(°C) \text{ kpm} \qquad (1)$$

For alcohols:

$$\Delta H_v = 6.0 + 0.055 t_b(°C) \text{ kpm} \qquad (2)$$

These equations appear reliable within a few tenths of a kilocalorie for most liquids boiling below about 200°C, but the relationship is not truly linear over a wider temperature range and larger errors result for higher boiling compounds.

Recently, I have been engaged in a study of cohesive energies in both liquids and solids. This is a complex and difficult area but certain useful information has already resulted which can appropriately be included here. More than 720 liquids are being studied. They include about 120 inorganic or organometallic compounds and more than 600 organic compounds. The inorganic compounds include at least five examples of each type: binary hydrogen, binary oxygen, binary sulfur, binary fluorine, binary chlorine, binary bromine, metal alkyls, oxyhalides, and a wide variety of miscellaneous compounds. The organic compounds include 157 alkanes, 90 alkenes, 52 cycloalkanes, 47 alkylbenzenes, 54 halides, 42 mercaptans and sulfides

---

[6] J. D. Cox and G. Pilcher, " Thermochemistry of Organic and Organometallic Compounds." Academic Press, New York, 1970 ; D. R. Stull, E. F. Westrum, Jr., and G. C. Sinke, " The Chemical Thermodynamics of Organic Compounds." Wiley, New York, 1969.

[7] I. Wadso, *Acta Chem. Scand.* **20**, 544 (1966).

30 nitrogen compounds, 32 alcohols, 22 esters, 18 ethers, 12 ketones, 7 aldehydes, and a large number of other functional and mixed-functional compounds. These compounds have boiling points that range from -56 to $380°C$ or over $436°$, and have a cohesive energy that ranges from 3.7 to 26.1 kpm or varies by a factor of 7.

Nearly 50 compounds having OH groups and therefore capable of protonic bridging are conspicuous for having cohesive energies appreciably higher than expected for their boiling points. All the rest appear to conform quite well to the empirical equation:

$$C_l = \frac{T_b}{68.3 - 0.0674\,T_b} \tag{3}$$

C_l is the cohesive energy in kpm and T_b the absolute boiling point. This is not an exact equation, but certainly suitable for estimating values in the calculation of standard heats of formation of liquid compounds in general from gaseous atomization energies. For about 98% of these compounds, the difference between calculated and experimental cohesive energies is less than 2 kpm, and for 93%, it is less than 1 kpm; 83% have differences of ± 0.5 kpm or lower, 66% being within ± 0.3 kpm. Wherever compounds differing by more than 1 kpm can be compared with other compounds of similar type, they appear exceptional, suggesting that the differences may originate with experimental error. Except for the alcohols and other OH compounds mentioned above, no *consistent* class deviations from Eq. (3) have been observed.

The cohesive energies of alcohols can be estimated quite well by adding 3.3 kpm to the value calculated from Eq. (3). For phenols, 1.9 kpm should be added. There is unquestionably a small structural effect on cohesive energy that is not quantitatively accounted for by change in boiling point, but in general this is less than 1 kpm and was adequately averaged out in the derivation of Eq. (3).

As might well be expected, the experimental cohesive energies of compounds that are crystalline solids at $25°$ show a more widely scattered relationship to boiling point. A similar study of 135 solid compounds, 41 inorganic and 94 organic, has produced the following approximate equation resembling Eq. (3):

$$C_s = \frac{T_b}{37.14 - 0.0172\,T_b} \tag{4}$$

Again, OH compounds are exceptional. Of the others, the average difference between calculated and experimental cohesive energies is about 1.6 kpm.

Obviously, intermolecular cohesive energies make an important contribution to heats of formation of both inorganic and organic liquids and solids. A fuller understanding of chemical bonding and reaction must therefore await further studies of cohesive energies as well as of the chemical bonds themselves.

Chapter 11

PREDICTING THE ENTHALPY OF REACTION

Experimental thermochemistry has provided invaluable data from which chemists can easily make reliable predictions about the stabilities of compounds and the directions of their reactions. Yet up to now these data have in a sense been "magic numbers." That is, the great utility of the experimental measurements has not included a very significant contribution to fundamental understanding. The methods of this book allow us to come a step closer to such understanding by suggesting not merely what the heat of formation, for example, should be, but also *why* it has the particular value which has been measured. We have become so accustomed to accepting and working with experimental data that we tend to take them for granted. Even when we appreciate the skillful experimentation essential to reliable measurement of such data, we are still inclined to concentrate our interest on our application of the data rather than the causes of their being.

At the risk of belaboring the obvious, therefore, this chapter includes a somewhat detailed analysis of a number of typical reactions, chosen to illustrate the application of bond energy calculations toward achieving a more fundamental understanding of chemical change.

Why Does Methane Burn?

The first example is taken largely from a recent paper,[1] which calls attention to the desirability of analyzing chemical change from a more fundamental viewpoint. The burning of methane was selected as a very common, well-known kind of chemical change which is accepted without question simply because everyone knows methane burns. It is indeed a simple matter, should the question of "Why?" arise, to look up the standard heats of formation of all the substances involved in this reaction and by

[1] R. T. Sanderson, *J. Chem. Educ.* **45**, 423 (1968).

simple arithmetic determine that the enthalpy change known as the heat of reaction is -191.8 kpm. In the gas phase reaction,

$$CH_4 + 2 O_2 \rightarrow CO_2 + 2 H_2O$$

3 moles of gaseous reactants form 3 moles of gaseous products. The state of disorder or randomness should be essentially the same in the products as in the reactants, which means that the entropy change should be very small. It is therefore not surprising to find that the free energy of the reaction is -191.4, nearly the same as the heat. In the familiar equation,

$$\Delta G^\circ = \Delta H^\circ - T \Delta S^\circ$$

in other words, $T \Delta S^\circ$ is almost negligibly small. Anybody knows that a high negative free energy change is associated with a spontaneity of reaction, so of course methane burns. Until recently it has not been possible to analyze the system in very much more detail than that.

One further step has of course been possible since it has long been recognized that the evolution of heat during a chemical change implies a total energy of association or bonding greater in the products than in the reactants. One can infer with confidence that, on the average, the bonds are stronger in the reactants than in the products. Therefore methane burns because the bonds in CH_4 and O_2, on the average, are not as strong as the bonds in CO_2 and H_2O. Formerly that was the end of the explanation. Now it is possible to say something useful about *why* the bonds in CO_2 and H_2O are, on the average, stronger.

There should be no need at this stage to repeat in detail all the bond energy calculations that pertain to the burning of methane. The results may be usefully tabulated as shown in Table 11-1. One may observe the before-and-after condition of each kind of atom, one by one. Carbon is more

TABLE 11-1

THERMOCHEMICAL DATA FOR THE OXIDATION OF METHANE[a]

Molecule	Bond	E_c (kpm)	E_i (kpm)	E/bond	$E_{calc.}$
CH_4	C—H	90.0	9.1	99.1	396.4
O_2	O=O	119.2	0	119.2	119.2
CO_2	C=O	119.1	73.0	192.1	384.2
H_2O	O—H	49.6	62.2	111.8	223.6

Reactants		Products	
4 C—H	396.4	2 C=O	384.2
2 O=O	238.4	4 O—H	447.2
Sum	634.8	Sum	831.4

[a] ΔH of reaction: calculated, -196.6; experimental, -191.8 kpm.

tightly held in methane than it finally is in carbon dioxide, by 12.2 kpm. There is no advantage to carbon in the burning of methane. Hydrogen becomes more tightly bound in water than it was to carbon, so here an advantage is gained. Finally, oxygen, despite its strong double bond in the O_2 molecule, becomes more strongly held by both hydrogen and carbon than it was in molecular oxygen. The overall advantage arises essentially from the bond polarity in the products which is nearly absent from the reactants. Notice that the polar energy is about 10% in methane and zero in O_2, whereas in CO_2 it constitutes 38% of the total energy and in water 56%.

It has been shown[1] that if it were not for the bond polarity in the products methane would not burn. The disadvantage to oxygen of using its O' energy in nonpolar water over the O'' energy in O_2 would make the *reactants* actually more stable. Again the great importance of electronegativity differences is illustrated.

Indeed, one may generalize that except in certain relatively nonpolar bonds where multiplicity effects can dominate, bond energies are usually higher in combinations of elements than in the pure elements because different elements are different in electronegativity.

The Synthesis of Ammonia

Again it would be easy to predict the combination of nitrogen and hydrogen to form ammonia, based on the exothermicity and negative free energy change for the reaction, as determined from tabulated data. But would one otherwise really be able to state with confidence that nitrogen atoms and hydrogen atoms would on the average be more tightly bound in ammonia molecules than they are in their very stable N_2 and H_2 molecules? After all, the N—N bond in N_2 is the strongest known bond except that in carbon monoxide, and of all homonuclear single bonds, hydrogen forms the strongest. Is the electronegativity difference between nitrogen and hydrogen sufficient to give to the ammonia molecule the necessary advantage of polarity?

First, what if there were no polarity in an N—H bond? The geometric mean of the N' energy, 39.2, and the H energy, 104.2, is 64.0 kpm. This would be the bond energy, which multiplied by 3 would give 192 kpm for the atomization energy of ammonia. The atomization energy of N_2 is 113.0 kpm of nitrogen atoms, and to produce three hydrogen atoms from their standard state as H_2 molecules would require another 156.3 kpm of ammonia. The total, 269.3 kpm, is far too high to encourage the formation of ammonia releasing only 192 kpm. The enthalpy of formation of ammonia would be the difference, +77.3 kpm.

In fact nitrogen has a sufficiently greater electronegativity than hydrogen to acquire a partial charge of −0.16, leaving each hydrogen atom with a partial charge of $+0.05$. This corresponds to $t_i = 0.11$. It also shortens the bond from the covalent radius sum of 1.06 to 1.02 Å. Although the hydrogen energy is reduced somewhat by the positive partial charge, the covalent contribution to the bond energy is still 57.6 kpm, not greatly lower than the nonpolar value of 64.0. But the ionic contribution is 35.8 kpm, bringing the total bond energy to $3 \times 93.4 = 280.2$ kpm. This is enough to atomize the necessary N_2 and H_2 and still be exothermic by -10.9 kpm. The loss of entropy associated with the change from 4 moles of gas to 2 comes close to eliminating this advantage to ammonia, but the free energy change for the synthesis is still about -4 kpm.

Gaseous Hydrolysis of Methyl Chloride

Another example of a chemical reaction which would not be easy to predict without quantitative data is the hydrolysis of methyl chloride to form methyl alcohol and hydrogen chloride:

$$CH_3Cl + H_2O \rightarrow CH_3OH + HCl$$

For methyl chloride, the C—H bond energy is calculated to be a total of 296.1. The C—Cl bond energy is calculated as 83.0 kpm. The atomization of $H_2O(g)$ requires 223.6 kpm, calculated. The total bond energy of the reactants is then 602.7 kpm.

In the products, the C—H bond energies total the same as for CH_3Cl, 296.1. The C—O bond energy is a little less than the C—Cl energy, 81.2 kpm. The bond energy of HCl is calculated as 103.7 kpm. The total bond energy in the products is thus 594.1 kpm. The enthalpy for the hydrolysis reaction is calculated to be the difference between 602.7 and 594.1, or $+8.3$. The experimental enthalpy of this reaction is $+7.2$, in good agreement with the calculated value. The gaseous hydrolysis is not favored, although the difference between reactants and products is relatively small.

It can be seen from the above data that the principal reason for the reluctance to hydrolyze comes from the substitution of one of the two O—H bonds in water by an H—Cl bond which is appreciably weaker. The C—O and the C—Cl bonds nearly balance one another. Then why is the H—Cl bond weaker than an O—H bond? The H—Cl nonpolar covalent bond energy parameter is 71.4 compared to only 55.8 for the H—O nonpolar energy. Two other factors favor the O—H bond, however. One is the shorter bond length, 0.96 compared to 1.27 Å. The other is greater polarity, with $t_i = 0.18$ compared to 0.16 for HCl. As a consequence of these differences, the covalent and ionic energies are: O—H, 49.6, 62.2; HCl, 61.9, 41.8 kpm. So the O—H bond is about 7 kpm stronger than the H—Cl bond. In the above reaction this makes the critical difference between reactants and products, and therefore in the probable direction of reaction. Indeed, the principal factor is the bond length, for if the bond in HCl were as short as in OH, the bond energy would be about 25 kpm greater than that of O—H. In that case, the reaction enthalpy would favor the hydrolysis.

Strontium Iodide and Lithium Fluoride

As a fourth illustration, let us consider a mixture of the solids strontium iodide and lithium fluoride. Would such a mixture be more stable or less stable than a mixture of strontium fluoride and lithium iodide? Table 11-2

TABLE 11-2

THERMOCHEMICAL DATA FOR LITHIUM HALIDE—STRONTIUM HALIDE[a]

$2 \text{LiF} + \text{SrI}_2 \rightarrow \text{SrF}_2 + 2 \text{LiI}$

$2 \text{LiF} + \text{SrI}_2 \rightarrow \text{SrF}_2 + 2 \text{LiI}$

Compound	Standard heat of formation (kpm)	
	Calc.	Exp.
LiF	− 150.5	− 146.5
LiI	− 68.7	− 64.6
SrF$_2$	− 286.5	− 290.3
SrI$_2$	− 131.8	− 135.5

[a] ΔH of reaction: calculated, + 8.9 kpm; experimental, + 9.0 kpm.

gives the calculated and experimental data. The LiF–SrI$_2$ mixture is seen to be more stable by about 9 kpm. The close agreement between calculated and experimental heats of formation, however, is fortuitous. It happens that each of the calculated values is about 4 kpm below the experimental value for the strontium compounds but 4 kpm above the experimental value for the lithium compounds. Thus the errors cancel. One would have greater basis for confidence in such calculations for solids if the calculated enthalpy of reaction were large enough to accommodate small errors.

Boron Trichloride and Zinc Fluoride

Finally, let us examine the possibility of a reaction between gas and solid:

$$2 \text{BCl}_3(g) + 3 \text{ZnF}_2(c) \rightarrow 2 \text{BF}_3(g) + 3 \text{ZnCl}_2(c)$$

The necessary data are given in Table 11-3. It will be seen that the chlorine atoms are more strongly bonded in BCl$_3$ than in ZnCl$_2$, but the fluorine atoms are more strongly bonded in BF$_3$ than in ZnF$_2$. The difference in the fluorides is greater by 191.6 kpm, whereas the difference in the chlorides is

TABLE 11-3

THERMOCHEMICAL DATA FOR ZINC FLUORIDE—BORON TRICHLORIDE[a]

Compound	t_i	E_c (kpm)	E_i (kpm)	Atomization energy (kpm)	
				Calc.	Exp.
ZnF_2	0.23	92.2	151.6	243.8	244.8
$ZnCl_2$	0.16	109.7	78.6	188.3	188.6
BCl_3	0.26	56.1	49.3	316.2	318.3
BF_3	0.36	62.4	91.9	462.9	463.0

$$2\,BCl_3 \;+\; 3\,ZnF_2 \;\longrightarrow\; 2\,BF_3 \;+\; 3\,ZnCl_2$$

$\Sigma\Delta Hf°$				
Calc.	− 188.8	− 526.8	− 543.4	− 297.3
Exp.	− 193.0	− 528.0	− 543.6	− 298.2

[a] $\Delta H°$ of reaction: − 125.1, calculated; 120.8, experimental.

only 70.8 kpm. Consequently the enthalpy of the reaction is − 120.8 kpm, experimental. The calculated value is − 125.1 kpm. There is no question about the feasibility of a metathetical reaction in the direction written above from the viewpoint of enthalpy.

A direct comparison of bond strengths is not possible in this kind of situation because of the difference between a bond in a molecule and bonding in the nonmolecular solid. In general, however, we can examine the causes of the stronger bonding of fluorine to boron than to zinc. First, the bond length is much shorter in BF_3, 1.30 compared to 2.04 Å. Second, since zinc and boron are very close together in electronegativity, one might expect them to have similar bond polarity; however, t_i is quite different in the two compounds. It is equal to the charge on fluorine in the ZnF_2, but in BF_3 it is the average of the charge on fluorine and the charge on boron which is three times greater. Thus t_i is only 0.23 in ZnF_2 but 0.36 in BF_3. The ionic energy contribution in ZnF_2 is 151.6 but in BF_3 it totals 275.7 kpm. Owing to both the shorter bond length and the much higher B—B energy compared to the Zn—Zn energy, the covalent energy is also much greater in BF_3, 187.2 compared to 92.2 in ZnF_2. Since the differences between the chlorides are much less, it is primarily the fluoride difference that is the dominant factor in determining the possible enthalpy of the reaction.

Summary

In principle, the possibility of understanding chemical bonds well enough to calculate their energies in hundreds of compounds provides a unique opportunity to gain insight into the enthalpies of thousands of chemical reactions. In practice, one would not necessarily wish to take time for a detailed analysis of each reaction of interest. Nevertheless, it should be reassuring to realize that tabulated standard heats of formation or heats of reaction are not merely numbers obtained from a calorimeter, but can be explained in a rational, quantitative manner from known properties of the component atoms.

As a result of the work described here, it is now possible to reach a fuller, although imperfect, understanding of a given reaction or a given compound than ever before. Perhaps it will be useful, as a suggestion to teachers of chemistry as well as an exercise to sharpen the intuitive processes of practicing chemists, to choose, for example, a compound of intermediate complexity and examine its formation and properties in detail. The purpose is not to emphasize the compound as such but to illustrate how any compound might be examined and how a useful chemical understanding of it may be attained.

The Burning of Carbon and the Properties of Carbon Dioxide

Let us consider the very familiar reaction,

$$C + O_2 = CO_2$$

and the nature of the bonding in the well-known product carbon dioxide.

The most stable state of carbon at one atmosphere and 25°C is graphite. When ignited in the presence of sufficient oxygen, graphite will burn, uniting with the oxygen to form the gaseous compound CO_2. The reaction proceeds spontaneously, emitting heat and light. From this observation we infer correctly that the bonds in a carbon dioxide molecule must be stronger than the average of the bonds which originally held the carbon in graphite and the oxygen in O_2 molecules. A quantitative analysis of the thermochemistry of this reaction should be instructive.

In a molecule of carbon dioxide the carbon atom is isolated from all other carbon atoms, whereas in graphite it was firmly bound to three close neighbors by strong bonds. The energy required to atomize graphite at 25° is experimentally determined as 171.3 kpm. It is not yet possible to explain why the graphite form is preferred over diamond, except from the experimental point of view that the latter is less stable by about 0.5 kpm. But we can rationalize the graphite energy by considering its known structure and calculating the bond energy. Graphite consists of parallel planes of condensed rings of carbon atoms. The interplanar distance is too great to involve other than van der Waals attractions, for which we have no present method of estimating the magnitude except that it should be relatively small. Within each plane, each carbon atom is equidistant from three other carbon atoms at 1.42 Å, the bond angles being 120°. Since each carbon atom uses four electrons to form only three bonds, each bond is of 1.33 order. This means 0.67 pi electron per bond. The empirical multiplicity factor for condensed aromatic ring type compounds is $1 + 0.33n$ where n is the average number of pi electrons per bond. For graphite then, $n = 0.67$ and the factor is 1.22. The bond energy is then calculated as:

$$E = \frac{83.2 \times 1.22 \times 1.54}{1.42} = 110.0 \text{ kpm}$$

To atomize graphite, since each bond broken liberates two atoms from that bond, breaking 1.5 bonds per carbon atom is sufficient for complete atomization, for which the energy is then $1.5 \times 110.0 = 165.0$. This allows about 6 kpm for the van der Waals energy between layers. If not exact, then this calculation shows approximately that the experimental atomization energy of graphite is consistent with the reasonable expectation based on our experience with bond energy calculations. Obviously carbon cannot form carbon dioxide unless 171.3 kpm is supplied to the graphite to separate its atoms.

Furthermore, in a molecule of carbon dioxide the carbon atom rests between two oxygen atoms which are thus separated completely from one another, whereas in the O_2 molecule they were held together with a bond energy of 119.2 kpm. The sum, $171.3 + 119.2 = 290.5$, represents the amount of energy required to be expended to produce enough separated atoms of carbon and oxygen to form one mole of CO_2 molecules. The observation that the burning of graphite releases energy means, therefore, that the energy released by combination of one mole of carbon atoms and two moles of oxygen atoms must exceed 290.5 kcal. Let us calculate the energy of the bonds in carbon dioxide to assure ourselves that this is so and to gain some insight as to the exact nature of the bonds.

First, the ability of a carbon atom to provide four half-filled outer orbitals for bonding is evident from its electronic configuration. Similarly, each atom of oxygen has two half-filled outer orbitals available for bonding. The most obvious combination of carbon with oxygen would thus involve two oxygen atoms per carbon atom, giving the empirical formula CO_2. Two possible modes of combination could correspond to this formula. One would be the monomeric molecule. The other would be the corresponding polymer, in which each carbon atom would be joined to four oxygen atoms by single bonds, and each oxygen atom to two carbon atoms. Since the former is the one experimentally observed, let us for the moment concentrate on it; we will come back to the other form later. To make full use of the bonding capacity of the carbon and oxygen in a single molecule, each oxygen atom must be attached to carbon through a double bond.

The electronegativity of oxygen, 5.21, is substantially higher than that of carbon, 3.79. We therefore predict that the bonds will have appreciable polarity, with oxygen negative and carbon positive. Polarity usually appears to produce bond shortening. Both polarity and multiplicity therefore contribute to closer equilibrium positions of the atoms than would be predicted from merely adding the nonpolar covalent radii of carbon and oxygen. There is not yet a completely satisfactory method of predicting the bond length quantitatively, but we can rely on the experimental measurement of 1.16 Å. As predicted, this is substantially shorter than the nonpolar covalent radius sum of 0.77 for carbon and 0.72 for oxygen, or 1.49 Å.

In a molecule of carbon dioxide, all the outer shell electrons of the carbon atom are involved in the bonding, but there are only two bonds. The farthest apart these two bonds can be, to minimize the interelectronic repulsions, is at opposite sides of the carbon atom. Therefore the molecule may be predicted to be linear, $O=C=O$, with a bond angle of 180°. This prediction is verified by experiment.

Now let us examine first qualitatively, then quantitatively, the polarity of the carbon to oxygen bonds. To simplify the picture, let us think only of a single covalent bond between carbon and oxygen at the beginning of its formation. In other words, picture a carbon atom and an oxygen atom adjacent to one another such that one valence electron of each atom, without distorting that atom in any way, also occupies a vacancy beside the valence electron of the other atom. The electronegativity of an atom may be regarded as representing the coulombic force between its nucleus and an electron from another atom which happens to be occupying an outermost vacancy of the first atom. The net attraction is then the charge on the electron times that fraction of the total nuclear charge which the electron can sense, divided by the square of the distance between them, which is the nonpolar covalent radius. The electron only senses a small fraction of the total nuclear charge because of the screening effect of the intervening electronic cloud. This fraction, called the "effective nuclear charge," is the difference between the real nuclear charge and the sum of the screening constants of the other electrons of its atom.

The coulombic force between the electron from the oxygen atom and the carbon nucleus is thus represented by the electronegativity of carbon, 3.79. The coulombic force between the electron from the carbon atom and the oxygen nucleus is represented by the electronegativity of oxygen, 5.21. The two electrons are initially subjected to attractive forces that are quite unequal. Since electrons of any atom or molecule are subjected only to the restraints placed upon them by the proximity of other electrons or nuclei, they are certainly free to move toward positions of minimum repulsion and maximum attraction, thus stabilizing the system. The two bonding electrons between the carbon and oxygen atom will therefore favor penetration toward the oxygen nucleus. Such penetration must have important effects on both atoms. The effect on the oxygen atom is first to increase the electron population around its nucleus. This increases the interelectronic repulsions, which in view of the unchanged nuclear charge must cause expansion of the electron cloud. Penetration of electrons toward the nucleus is in effect reduced, and the screening of the nuclear charge increases. These changes correspond to a reduction of the oxygen electronegativity, since an electron at the periphery of oxygen is now farther from the nucleus and senses a reduced effective nuclear charge, both of which must decrease the coulombic force between the electron and the nucleus.

The effect on the carbon atom is first to decrease the electron population around its nucleus. This diminishes the interelectronic repulsions, allowing the electrons to be drawn closer, on the average, to the nucleus. The effective nuclear charge increases because of the reduced screening effects. Both the reduced radius of the cloud and the increased effective nuclear charge increase the coulombic force between an electron at the periphery and the nucleus. In other words, the electronegativity increases.

By uneven sharing of the bonding electrons, the electronegativity of the initially more electronegative oxygen decreases while the electronegativity of the initially less electronegative carbon increases. According to the principle of electronegativity equalization, this adjustment of each atom ceases when the electronegativities of all the atoms of the molecule have attained the same intermediate value. The corollary to this principle is that the intermediate value which is reached is the geometric mean of the initial electronegativities of all the atoms. We can now begin to take a more quantitative view of this equalization process. The geometric mean for CO_2 is the cube root of the electronegativity product, $3.79 \times 5.21 \times 5.21$, which is 4.69. The electronegativity of carbon has increased from 3.79 to 4.69, or by 0.90. The electronegativity of oxygen has decreased from 5.21 to 4.69, or by 0.52. These changes are the result of uneven sharing which places a partial negative charge on the oxygen, since the bonding electrons spend more than half time more closely associated with the oxygen nucleus. A partial positive charge remains on the carbon. During the shorter time the bonding electrons spend around the carbon, therefore, they are attracted just as strongly as during the longer time they spend around the oxygen nucleus. Even attraction results from uneven sharing.

The partial charge on a combined atom is defined as the ratio of the change in electronegativity from free atom to compound to the change in electronegativity from free atom to (plus or minus) unit-charged ion. Values for the latter are available for many of the elements, and for carbon and oxygen they are 4.05 and 4.75. The partial charge on carbon is then $0.90/4.05 = 0.22$. The partial charge on oxygen, negative to correspond to the loss in electronegativity, is $0.52/4.75 = -0.11$.

Up to this point, our description of the CO_2 molecule includes its molecular formula, its structure in terms of bond angle and bond length, its bond type, which is polar double bond, and a quantitative evaluation of the bond polarity in terms of the partial charges on the combined atoms. We may now proceed with calculation of the bond energy, for most practical purposes thus completing the description of carbon dioxide.

The standard procedure described here is to treat the very complex system that is the carbon dioxide molecule as though the bonds were completely nonpolar covalent part of the time and completely ionic the rest of the time. That is, although the molecule may be properly represented in terms of molecular orbitals, we depict its bonds as the sum of components that are separable (for purposes of calculation only).

One is the nonpolar covalent component, which would be represented by the even electron sharing between carbon and oxygen as though no difference in electronegativity existed. The nonpolar bond energy, at least approximately, may be regarded as the coulombic energy of attraction between the bonding electrons and the shielded nucleus of each atom. Each atom is assumed to contribute to the nonpolar covalent bond energy of the heteronuclear bond exactly as it would contribute to a single homonuclear bond. Therefore the nonpolar covalent energy of the heteronuclear bond is taken as the geometric mean of the two homonuclear single covalent bond energies. Correction must be made, however, for variation of the bond length from the nonpolar covalent radius sum, for shorter bonds are stronger and longer bonds weaker. The correction factor is R_c/R_0, where R_c is the sum of the two nonpolar covalent radii, and R_0 is the experimentally measured bond length. The nonpolar covalent bond energy, E_c, is then

$$E_c = \frac{R_c(E_{AA}E_{BB})^{\frac{1}{2}}}{R_0}$$

For the carbon–oxygen double bond, the geometric mean of the homonuclear single bond energies for carbon, 83.2, and oxygen, 66.7, is 74.5. The oxygen value, 66.7, was derived from a study of the O_2 molecule. It differs from the usually accepted single bond energy of 34.0 because this lower value represents the full weakening power of the lone pair(s) whereas formation of a double bond partly removes this weakening power.

The other component of the bond is the ionic component, E_i, consisting of the electrostatic energy of attraction between unit opposite charges assumed to be located at the nuclei of the two atoms:

$$E_i = \frac{332e^2}{R_0}$$

The factor 332 converts the energy to kilocalories per mole (kpm). A double bond is stronger than a single bond at the same bond length by the factor 1.50, as determined from a study of the thermochemistry of alkenes. Both the nonpolar covalent and the ionic contributions to the bond energy in carbon dioxide must therefore be multiplied by 1.50.

Finally, before we can calculate the covalent and ionic energy contributions to the bond, we much evaluate the ionic weighting coefficient, t_i. This is a crucial part of the calculation, but easily performed, for t_i is simply the average of the partial charges 0.11 and 0.22, or 0.17. Then t_c the covalent weighting coefficient is $1.00 - 0.17 = 0.83$.

The covalent contribution to the carbon–oxygen bond is

$$E_c = \frac{0.83 \times 74.5 \times 1.50 \times 1.49}{1.16} = 119.1 \text{ kpm}$$

The ionic energy contribution is

$$E_i = \frac{0.17 \times 332 \times 1.50 =}{1.16} 73.0 \text{ kpm}$$

The sum, 192.1 kpm, is the energy per carbon–oxygen bond. The total atomization energy of carbon dioxide is twice this value, or 384.2 kpm. It does indeed exceed the 290.5 kpm total atomization energy of the individual elements, mentioned earlier, as predicted. For the first time, by this method, we have a quantitative evaluation of the relative contributions of covalence and ionicity to the total energy of a bond. The covalent energy is 62% of the total bonding energy, and the ionic energy contributes 38%.

Returning now to the atomization energies of carbon and oxygen and their effect on the heat evolved when carbon burns, we can calculate that the net heat or reaction enthalpy must be equal to $290.5 - 384.2 = -93.7$ kpm. This is the calculated standard heat of formation of carbon dioxide. The experimental value is -94.1 kpm, in excellent agreement with the calculated value. In consideration of the several debatable assumptions on which the calculation is based, one might well feel obliged to concede that such excellent agreement could be fortuitous if it were observed for just the one compound. Taken together with similar agreement for hundreds of other compounds, however, the results for carbon dioxide lend strong support to the conclusion that all the required assumptions must be essentially correct.

There is one more point to be considered. That is the question of why carbon and oxygen choose the monomeric molecule rather than the polymer as their most stable combined state. For in general, are not four single bonds likely to be more stable than two double bonds? Indeed they are. In fact, if the only difference between monomer and polymer were bond multiplicity and bond length, one could readily calculate, assuming a reasonable single bond length of 1.45 Å and, of course, no multiplicity factor of 1.50, that the energy of a single bond in the polymer would be 102.5 kpm. Four such bonds per carbon atom must be broken for atomization of the polymer, so the total atomization energy would be 410.0 kpm, or 26 kpm greater than for the monomer. The stable form of carbon dioxide would then indeed be the polymeric solid, not the gas.

However, this conclusion overlooks an important fact. The homonuclear single covalent bond energy of oxygen, without the reduction of lone pair weakening made possible by formation of a double bond, is only 34.0 instead of 66.7 kpm. Then the geometric mean bond energy is only 53.2 instead of 74.5. For the carbon–oxygen bond in the polymer, therefore, the covalent energy contribution is

$$E_c = \frac{0.83 \times 53.2 \times 1.49}{1.45} = 45.4 \text{ kpm}$$

The ionic energy contribution is

$$E_i = \frac{0.17 \times 332}{1.45} = 38.9 \text{ kpm}$$

The sum, 84.3 kpm for the single bond energy, must be multiplied by 4 to give the total atomization energy of the polymer, which is only 337.2 kpm compared to 384.2 for the monomer. The difference, +47.0 kpm, is the calculated enthalpy of polymerization of carbon dioxide. Coupled with the undoubted decrease in entropy that must accompany polymerization to a solid from the gas, this ensures that a high positive free energy change would be required in order for carbon dioxide gas to polymerize.

To summarize, the burning of carbon to form carbon dioxide gas has been completely and quantitatively rationalized on the basis of a knowledge of the following data: the atomic structure, nonpolar covalent radius, electronegativity, and homonuclear bond energy of both carbon and oxygen, and the bond length in carbon dioxide. The chemical formula, molecular structure, bond polarity, and bond energy have all been shown to be the reasonable if not inevitable results of the inherent nature of the separate atoms. The establishment of a quantitative cause-and-effect relationship between the qualities of atoms and the properties of their compounds is a highly desirable step toward a practical understanding of chemistry.

APPENDIX

Miscellaneous Notes on Table A

Unless otherwise indicated, all substances are gaseous. See Chapter 9 for data on nonmolecular solids. If the bond is discussed within the main text, no further comments are given here.

H_2Te, HI

For both of these molecules, the calculated bond energy is too low by what seems to be a significant amount, but this is corrected by adopting the Te'' and I'' homonuclear energies. There is really no firm justification for using these double prime energies, other than to produce better agreement, unless there is something unusual about these bonds. Their most unusual aspect is that they represent greater differences in atomic radius than any other compounds listed. One may speculate that the lone pair weakening effect may here be somehow reduced by the fact that the hydrogen atom is so small relative to the Te or I atom. Otherwise one might expect the bond to be weakened rather than strengthened by the disparity of size of atoms.

M2 GASEOUS OXIDES

It is noteworthy that to the very limited extent of the available experimental data, the gaseous MO molecules appear to have single bonds, in keeping with the concept of a second sp orbital of the metal being directed opposite to the first.

M2 HALIDES

The experimental data for the gaseous monohalides of the M2 elements are probably not sufficiently reliable to justify any extensive rationalization of the choice of homonuclear bond energy parameters. For the dihalides,

171

the choice is dictated by the agreement obtained between calculated and experimental atomization energies. Unfortunately no more fundamental rationalization is yet available.

$B(CH_3)_3$.

The problem of the C—H bond energy in organometallic compounds is recognized but remains to be solved. There is preliminary evidence that considerable variation in C—H energy may occur in these compounds. This may be the origin of the lack of agreement here.

BN, BO, BS,

The assignments of bond type here are probably not to be taken too seriously until more confidence in the experimental energies is justified.

$B(OH)_3$, BF_2OH, $BF(OH)_2$

It is possible that internal protonic bridging in these gaseous molecules may add to the atomization energy. This could amount to about 15 kpm for $B(OH)_3$, about 5 kpm for BF_2OH, and perhaps about 10 kpm for $BF(OH)_2$, giving somewhat better agreement between calculated and experimental values.

B_2F_4, B_2Cl_4

For both these molecules, the calculated atomization energy is significantly lower than the experimental value. For each, the B—B energy was calculated as reduced by mutual positive charge repulsion. Even without this correction, the calculated values would be too low. This suggests some additional energy may be derived from electrostatic attractions between positive boron and the negative halogen on the adjacent atom.

$B_3N_3H_6$, $B_3N_3H_3Cl_3$

The assignment of B—N″ bonds to the chloride but an average of B—N′ and B—N″ to the borazene itself seems inconsistent. However, this may be rationalized to some degree by recognizing that the partial charge on boron is only 0.19 in borazene but 0.27 in the chloride.

AlO

The single bond seems to be consistent with other examples of diminished multiplicity between atoms unlike in size.

THALLIUM (I) HALIDES

The difference between experimental and calculated bond energies (in kpm) is TlF, 13.9; TlCl, 13.1; TlBr, 13.2; and TlI, 14.1; averaging 13.6.

Experimental bond energies of the monohalides of boron and aluminum are similarly consistently higher than the calculated values. Perhaps the 13.6 reflects a difference in bonding orbitals, such that sp^2 is less stable than s^2p by 13.6 kpm of orbitals.

COX_2

The conventional picture of a carbonyl halide such as $O{=}CX_2$ appears inadequate. The C—O bond seems better represented as an average of an ordinary double bond $C{=}O''$ and a coordinate covalent bond, $C{-}O'''$. This is another problem that needs solving.

AsX_3, SbX_3

The experimental atomization energy of AsF_3 seems much too high for any reasonable atomic properties of arsenic and assignment of bond type. Experimental atomization energies of $AsCl_3$ and $AsBr_3$ also appear high, although only by about 4%. Similar results are obtained for trihalides of antimony, and again without satisfactory explanation.

SF_6, SF_5Cl, SeF_6, TeF_6

The calculated atomization energies of the first two compounds appear too high, but the energies of the second two compounds are too low. These are the only examples here of M6 compounds requiring promotion of the final s^2 lone pair. Possibly the deficit in the first two is related to promotional energy. Alternatively it may reflect steric bond weakening through crowding, which is more consistent with a 36.6 kpm deficiency for SF_5Cl and only a 19 kpm deficiency for SF_6, and no deficiency for either SeF_6 or TeF_6 where the central atom is larger.

SULFUR OXYHALIDES

The assignment of S'—O''' bond type seems appropriate for SOF_2, $SOCl_2$, SO_2F_2, but S'—O'' appears more suitable for SO_2Cl_2 and $SOBr_2$. The charges on S and O are SOF_2, 0.25, -0.01; SO_2F_2, 0.25, -0.01; $SOCl_2$, 0.16, -0.09; $SOBr_2$, 0.11, -0.13; SO_2Cl_2, 0.18, -0.07. There is no obvious consistency; this is another area where further study is needed.

ClO_3F, $HClO_4$

It is interesting that in both these compounds, the Cl—O' bond appears to represent reality, with no evidence of multiplicity or reduced weakening except bond shortening, even though these must be coordinate covalent bonds. The shortening is therefore noteworthy. It appears that a coordinate covalent Cl—O bond is shorter than an ordinary covalent bond as in Cl_2O by about 0.2 Å.

TABLE A

ATOMIZATION ENERGY DATA AND
HEATS OF FORMATION OF INORGANIC COMPOUNDS

Compound Bond	$E(AB)$	R_o	δ_A	δ_B	t_i	E_c	E_i	E_{at} (kpm) Calc.	Exp.	$\Delta Hf°$ (kpm) Calc.	Exp
OH											
H—O	53.6	1.03	0.19	−0.19	0.19	43.8	61.2	105.0	102.4	+6.7	+9.
H₂O											
H—O	55.8	0.96	0.12	−0.25	0.18	49.6	62.2	223.6	221.6	−59.8	−57.
H₂O₂											
H—O	53.6	0.97	0.19	−0.19	0.19	46.5	65.0	223.0			
O—O	34.0	1.49	−0.19	−0.19	0	32.9	0	32.9			
								255.9	256.0	−32.5	−32.
HOF											
H—O	49.8	0.96	0.30	−0.10	0.20	43.2	69.2	112.4			
O—F	35.9	1.40	−0.10	−0.20	0.05	34.8	11.8	46.6			
								159.0	156.7	−28.4	−26.
HS											
H—S	73.2	1.36	0.07	−0.07	0.07	68.1	17.1	85.2	87.1	+33.5	+34.
H₂S											
H—S	73.9	1.33	0.05	−0.09	0.07	70.3	17.5	175.6	175.7	−4.8	−4.
H₂Se											
H—Se	66.1	1.47	0.05	−0.11	0.08	61.2	18.1	158.6	146.3	−5.2	+7.
H₂Te											
H—Te″	63.0	1.66	0	0	0	63.4	0	126.8	125.9	+22.9	+23.
HF											
H—F	54.4	0.92	0.25	−0.25	0.25	45.7	90.2	135.9	135.8	−64.9	−64.
HCl											
H—Cl	71.4	1.27	0.16	−0.16	0.16	61.9	41.8	103.7	103.3	−22.5	−22.
HBr											
H—Br	65.1	1.41	0.12	−0.12	0.12	59.3	28.3	87.6	87.5	−8.8	−8.
HI											
H—I′	60.1	1.61	0.04	−0.04	0.04	59.1	8.2	67.3	71.3	+10.3	+6.
H—I″	63.2	1.61	0.04	−0.04	0.04	62.2	8.2	70.4	71.3	+7.2	+6.
LiH(g)											
Li—H	54.4	1.60	0.49	−0.49	(0)	56.4	0	56.4	59.8	+34.1	+30.
LiH(c)											
Li—H	54.4	2.04	0.49	−0.49	(0)	44.3	0	132.9	112.1	−42.4	−21.
LiOH											
Li—O	30.0	1.82	0.90	−0.60	0.75	8.5	136.8	145.3			
H—O	59.5	0.98	−0.30	−0.60	(0)	63.2	0	63.2			
								208.5	208	−58.4	−58
LiF											
Li—F	32.7	1.55	0.74	−0.74	0.74	11.2	158.5	169.7	136.8	−112.4	−79.
LiCl											
Li—Cl	40.6	2.03	0.65	−0.65	0.65	16.3	106.3	122.6	114.3	−55.1	−46.

(continued

ABLE A—*continued*

Compound Bond	$E(AB)$	R_o	δ_A	δ_B	t_i	E_c	E_i	E_{at} (kpm)		$\Delta Hf°$ (kpm)	
								Calc.	Exp.	Calc.	Exp.
Br Li—Br	36.2	2.17	0.61	−0.61	0.61	16.1	93.3	109.4	101.9	−44.3	−36.8
I Li—I	32.0	2.39	0.53	−0.53	0.53	16.8	73.6	90.4	85.7	−26.5	−21.8
aH(g) Na—H	43.3	1.89	0.50	−0.50	(0)	42.6	0	42.6	48.1	+35.4	+29.9
aH(c) Na—H	43.3	2.44	0.50	−0.50	(0)	33.1	0	99.3	91.7	−21.2	−13.7
aF Na—F	26.1	1.85	0.75	−0.75	0.75	7.9	134.6	142.5	114.8	−97.8	−70.1
aCl Na—Cl	32.4	2.36	0.67	−0.67	0.67	11.5	94.3	105.8	99.0	−50.9	−44.1
aBr Na—Br	28.8	2.50	0.62	−0.62	0.62	11.7	82.3	94.0	88.8	−41.5	−36.3
aI Na—I	25.5	2.71	0.54	−0.54	0.54	12.4	66.2	78.6	72.2	−27.3	−20.9
H(g) K—H	37.1	2.24	0.60	−0.60	(0)	37.8	0	37.8	43.5	+35.7	+30.0
H(c) K—H	37.1	2.85	0.60	−0.60	(0)	29.7	0	89.1	88.0	−15.6	−14.5
F K—F	22.4	2.14	0.85	−0.85	0.85	4.2	131.9	136.1	118.1	−95.9	−77.9
Cl K—Cl	27.8	2.67	0.76	−0.76	0.76	7.4	94.5	101.9	101.6	−51.5	−51.2
Br K—Br	24.7	2.82	0.71	−0.71	0.71	7.9	83.6	91.5	91.0	−43.5	−43.0
I K—I	21.8	3.05	0.63	−0.63	0.63	8.7	68.6	77.3	77.8	−30.5	−31.0
bH(g) Rb—H	35.9	2.37	0.63	−0.63	(0)	37.6	0	37.6	38.7	+34.1	+33
bH(c) Rb—H	35.9	2.98	0.63	−0.63	(0)	30.0	0	90.0	92.0	−18.3	−20.3
bF Rb—F	21.7	2.24	0.86	−0.86	0.86	4.0	127.5	131.5	115.7	−93.0	−77.2
bCl Rb—Cl	26.9	2.79	0.78	−0.78	0.78	6.7	92.8	99.5	100.5	−50.8	−51.8
bBr Rb—Br	23.9	2.95	0.73	−0.73	0.73	7.2	82.2	89.4	90.4	−43.1	−44.1
bI Rb—I	21.2	3.18	0.65	−0.65	0.65	8.1	67.9	76.0	76.8	−30.9	−31.7
sH(g) Cs—H	33.4	2.49	0.65	−0.65	(0)	35.8	0	35.8	41.9	+35.1	+29
sH(c) Cs—H	33.4	3.19	0.65	−0.65	(0)	28.0	0	84.0	82.	−13.2	−11

(*continued*)

TABLE A—*continued*

Compound Bond	E(AB)	R_o	δ_A	δ_B	t	E_c	E_i	E_{at} (kpm) Calc.	Exp.	$\Delta Hf°$ (kpm) Calc.	Exp
CsF											
Cs—F	20.1	2.35	0.90	−0.90	0.90	2.6	127.1	129.7	116.2	−92.1	−78
CsCl											
Cs—Cl	25.0	2.91	0.81	−0.81	0.81	5.5	92.4	97.9	101.	−50.1	−53
CsBr											
Cs—Br	22.2	3.07	0.77	−0.77	0.77	5.8	83.3	89.1	91.	−43.7	−46
CsI											
Cs—I	19.7	3.32	0.69	−0.69	0.69	6.8	69.0	75.8	75.	−31.6	−31
BeH											
Be—H	80.5	1.34	0.16	−0.16	(0)	73.9	0	73.9	53		
BeO											
Be—O′	46.1	1.60	0.35	−0.35	0.35	30.5	72.6	103.1	106.9	+34.8	+31
BeF											
Be—F‴	79.1	(1.41)	0.41	−0.41	0.41	53.6	96.5	150.1	146.9	−52.9	−49.
BeF₂											
Be—F‴	79.1	1.43	0.58	−0.29	0.44	50.2	102.2	304.8	303.6	−188.7	−187.
BeCl											
Be—Cl‴	70.0	1.75	0.32	−0.32	0.32	51.7	60.7	112.4	103.7	−5.0	+3.
BeCl₂											
Be—Cl‴	70.0	1.77	0.46	−0.23	0.34	49.6	63.8	226.8	222.8	−90.3	−86.
BeBr											
Be—Br‴	60.1	1.84	0.28	−0.28	0.28	48.2	50.5	98.7	122?	+6.3	−17.0
BeBr₂											
Be—Br‴	60.1	1.90	0.39	−0.20	0.29	46.0	50.7	193.4	190.6	−61.7	−58.
BeI											
Be—I‴	52.3	2.05	0.20	−0.20	0.20	45.7	32.4	78.1	102.8?	+25.7	+1.0
BeI₂											
Be—I‴	52.3	2.12	0.27	−0.14	0.21	43.6	32.9	153.0	152.3	−23.7	−23.
MgF											
Mg—F″	48.9	1.75	0.55	−0.55	0.55	26.3	104.3	130.6	132.0	−68.7	−70.
MgF₂											
Mg—F′	36.2	1.77	0.81	−0.41	0.61	16.7	114.4	262.2	258.9	−181.4	−178.
MgCl											
Mg—Cl′	44.8	(2.18)	0.47	−0.47	0.47	25.8	71.6	97.4		−25.3	
MgCl₂											
Mg—Cl′	44.8	2.18	0.68	−0.34	0.51	23.9	77.7	203.2	204.3	−102.0	−103.
MgBr											
Mg—Br′	40.0	2.34	0.42	−0.42	0.42	25.0	59.6	84.6	82.	−14.9	−13
MgBr₂											
Mg—Br′	40.0	2.34	0.61	−0.31	0.46	23.3	65.3	177.2	173.4	−80.8	−77.(
CaO											
Ca—O′	34.1	1.82	0.57	−0.57	0.57	19.8	104.0	123.8	93	−21.7	+9
CaF											
Ca—F″	48.6	2.02	0.62	−0.62	0.62	22.4	101.9	124.3	127	−62.9	−66

(*continued*

ABLE A—*continued*

Compound Bond	$E(AB)$	R_o	δ_A	δ_B	t_i	E_c	E_i	E_{at} (kpm) Calc.	E_{at} (kpm) Exp.	$\Delta Hf°$ (kpm) Calc.	$\Delta Hf°$ (kpm) Exp.
F$_2$ Ca—F″	48.6	(2.02)	0.93	− 0.47	0.70	17.7	115.0	265.4	268.3	− 185.1	− 188
F Sr—F″	47.5	(2.10)	0.66	− 0.66	0.66	20.1	104.3	124.4	127	− 66.5	− 69
F$_2$ Sr—F″	47.5	(2.10)	1.01	− 0.51	0.76	14.2	120.2	268.8	263	− 192.0	− 186
O Ba—O′	29.1	1.94	0.67	− 0.67	0.67	13.4	114.7	128.1	130	− 26.5	− 28
F Ba—F″	41.5	2.17	0.73	− 0.73	0.73	13.9	111.7	125.6	136	− 64.7	− 75
F$_2$ Ba—F″	41.5	2.17	1.12	− 0.56	0.84	8.2	128.5	273.4	273	− 193.6	− 193
F$_2$ Zn—F′	40.5	1.81	0.45	− 0.23	0.34	29.7	62.3	184.0	183.8	− 115.2	− 115.0
Cl$_2$ Zn—Cl$_2$	50.2	(2.09)	0.32	− 0.16	0.24	41.8	38.1	159.8	157.2	− 70.6	− 68
Br$_2$ Zn—Br′	44.8	(2.24)	0.26	− 0.13	0.20	39.1	29.6	137.4	131.8	− 53.0	− 47.4
I$_2$ Zn—I′	39.6	2.42	0.14	− 0.07	0.11	38.3	15.1	106.8	100.	− 24.8	− 18
I B—H	84.6	1.23	0.08	− 0.08	(0)	78.4	0	78.4	79.1	+ 108.2	+ 107.5
I$_3$ B—H	84.5	(1.14)	0.13	− 0.04	(0)	84.5	0	253.5	266.8	+ 37.3	+ 24
H$_6$ B—H	84.5	1.19	0.13	− 0.04	0.09	73.7	25.1	395.2			
		1.33	0.13	− 0.04	0.09	65.9	22.5	176.8	(4 half-bonds)		
								572.0	573.1	+ 9.6	+ 8.5
I(OCH$_3$)$_2$ B—H	82.4	(1.10)	0.23	0.05	0.09	77.7	27.2	104.9			
C—H	(98.7)							592.2			
C—O′	53.2	(1.43)	0.01	− 0.31	0.16	46.6	37.1	167.4			
B—O″	67.6	(1.38)	0.23	− 0.31	0.27	55.1	65.0	240.2			
								1104.7	1099.9	− 143.3	− 138.5
CH$_3$)$_3$ C—H	(98.7)							888.3			
B—C	75.5	1.56	0.17	− 0.06	0.12	67.8	25.5	279.9			
								1168.2	1147.0	− 50.9	− 29.7
N B≡N	80.5	1.28	0.20	− 0.20	0.20	138.9	91.9	230.8	92.7	+ 16.7	+ 154.8

(*continued*)

TABLE A—*continued*

Compound Bond	$E(AB)$	R_o	δ_A	δ_B	t_i	E_c	E_i	E_{at} (kpm) Calc.	Exp.	ΔHf^o (kpm) Calc.	Exp.
BO											
B≡O″	67.6	1.20	0.27	−0.27	0.27	95.1	112.1	207.2⎫	(average)		
B—O‴	82.5	1.20	0.27	−0.27	0.27	77.3	74.7	152.0⎭			
							Average	179.6	188	+14.5	+6
B₂O₃											
B—O″	67.6	1.36	0.34	−0.23	0.28	55.2	68.3	247.0			
B≡O″	67.6	1.24	0.34	−0.23	0.28	90.8	112.4	406.4			
								653.4	649.5	−205.6	−201
HOBO											
H—O	55.2	(0.96)	0.14	−0.23	0.19	48.5	65.7	114.2			
B—O″	67.6	(1.38)	0.33	−0.23	0.28	54.3	67.4	121.7			
B≡O″	67.6	(1.24)	0.33	−0.23	0.28	90.7	112.6	203.3			
								439.2	440.1	−133.4	−134
B(OH)₃											
H—O	55.5	(0.96)	0.13	−0.24	0.19	48.7	65.7	343.2			
B—O″	67.6	(1.38)	0.32	−0.24	0.28	54.3	67.4	365.1			
								708.3	728.2	−217.7	−237
B(OCH₃)₃											
H—C	(98.7)							888.3			
C—O′	53.2	1.43	0.01	−0.29	0.15	47.1	34.8	245.7			
B—O″	67.6	1.38	0.25	−0.29	0.27	55.1	65.0	360.3			
								1494.3	1511.1	−198.2	−215
BS											
B≡S‴	68.7	1.61	0.15	−0.15	0.15	101.1	46.5	147.6	119.4	+53.5	+81
BF											
B—F‴	82.9	(1.30)	0.33	−0.33	0.33	65.4	84.3	149.7	181	+3.7	−28
BF₂											
B—F‴	82.9	1.30	0.46	−0.23	0.34	64.4	86.8	302.4	302.3	−130.1	−130
BF₃											
B—F‴	82.9	1.30	0.53	−0.18	0.36	62.4	91.9	462.9	463.0	−271.7	−271
B₂F₄											
B—B	68.4	1.67	0.46	0.46	0	36.2	0	36.2			
B—F‴	82.9	1.32	0.46	−0.23	0.34	63.4	85.5	495.6			
								631.8	688.8		−344
BOF											
B—F‴	82.9	1.30	0.42	−0.26	0.34	64.5	86.8	151.3			
B≡O″	67.6	(1.23)	0.42	−0.16	0.29	90.2	117.4	207.6			
								358.9	358	−145.9	−145
BF₂OH											
H—O	52.0	(0.96)	0.24	−0.16	0.20	45.1	69.4	114.5			
B—O″	67.6	(1.38)	0.43	−0.16	0.29	53.6	69.7	123.3			

(*continued*)

BLE A—*continued*

mpound Bond	E(AB)	R_o	δ_A	δ_B	t_i	E_c	E_i	E_{at} (kpm) Calc.	Exp.	$\Delta Hf°$ (kpm) Calc.	Exp.
3—F'''	82.9	(1.30)	0.43	−0.26	0.34	64.3	86.8	302.2			
								540.0	544.9	−256.0	−260.9
(OH)$_2$											
1—O	54.3	(0.96)	0.17	−0.21	0.19	47.7	65.7	226.8			
3—O''	67.6	(1.38)	0.37	−0.21	0.29	53.6	69.8	246.8			
3—F'''	82.9	(1.30)	0.37	−0.30	0.33	65.3	84.3	149.6			
								623.2	626.5	−246.4	−249.7
$_2$Cl											
3—Cl'''	73.3	(1.75)	0.49	−0.06	0.27	55.3	51.2	106.5			
3—F'''	82.9	(1.30)	0.49	−0.22	0.35	63.4	89.3	305.4			
								411.9	414.2	−210.5	−212.8
Cl$_2$											
3—Cl'''	73.3	(1.75)	0.44	−0.10	0.27	55.3	51.2	213.0			
3—F'''	82.9	(1.30)	0.44	−0.25	0.35	63.4	89.3	152.7			
								365.7	365.8	−154.1	−154.2
'l											
3—Cl'''	73.3	(1.76)	0.24	−0.24	0.24	57.3	45.3	102.6	120.4	+61.0	+42.
l$_2$											
3—Cl'''	73.3	1.73	0.34	−0.17	0.25	57.5	48.0	211.0	213.4	−18.3	−20.7
l$_3$											
3—Cl'''	73.3	1.75	0.39	−0.13	0.26	56.1	49.3	316.2	318.3	−94.4	−96.5
Cl$_4$											
3—B	68.4	(1.75)	0.34	0.34	0	42.3	0	42.3			
3—Cl'''	73.3	(1.73)	0.34	−0.17	0.25	57.5	48.0	422.0			
								464.3	502.6		−117.2
Cl											
3—Cl'''	73.3	(1.75)	0.36	−0.15	0.26	56.1	49.3	105.4			
3=O''	67.6	(1.24)	0.36	−0.21	0.29	89.4	116.4	205.8			
								311.2	298.2	−88.0	−75
($_3$O)$_2$BCl											
C—H	(98.7)							592.2			
C—O'	53.2	(1.43)	0.02	−0.28	0.15	47.1	34.8	163.8			
3—O''	67.6	(1.38)	0.26	−0.28	0.27	55.1	65.0	240.2			
3—Cl'''	73.3	(1.75)	0.26	−0.23	0.25	56.8	47.4	104.2			
								1100.4	1116.2	−162.4	−178.2
H$_5$O)$_2$BCl											
C—H	(98.7)							987.0			
C—C	83.2	(1.54)	0	0	0	83.2	0	166.4			
C—O'	53.2	(1.43)	0	−0.30	0.15	47.1	34.8	163.8			
3—O''	67.6	(1.38)	0.24	−0.30	0.27	55.1	65.0	240.2			

(*continued*)

TABLE A—*continued*

Compound Bond	$E(AB)$	R_o	δ_A	δ_B	t_i	E_c	E_i	E_{at} (kpm) Calc.	E_{at} (kpm) Exp.	$\Delta Hf°$ (kpm) Calc.	$\Delta Hf°$ (kpm) Exp.
B—Cl'''	73.3	(1.75)	0.24	−0.25	0.24	57.7	45.5	103.2			
								1660.6	1685.0	−171.6	−196
$(C_2H_5O)BCl_2$											
C—H	(98.7)							493.5			
C—C	83.2	(1.54)	0.02	0.02	0	81.6	0	81.6			
C—O'	53.2	(1.43)	0.02	−0.28	0.15	47.1	34.8	81.9			
B—O''	67.6	(1.38)	0.27	−0.28	0.27	55.1	65.0	120.1			
B—Cl'''	73.3	(1.75)	0.27	−0.23	0.25	56.8	47.4	208.4			
								985.5	1004.5	−130.1	−149
BBr											
B—Br'''	63.0	(1.87)	0.20	−0.20	0.20	52.8	35.5	88.3	99.6	+72.9	+61
BBr_2											
B—Br'''	63.0	1.87	0.27	−0.14	0.21	52.2	37.3	179.0	176.8	+8.9	+11
BBr_3											
B—Br'''	63.0	1.87	0.31	−0.10	0.21	52.2	37.3	268.5	264.3	−53.9	−49
BI											
B—I'	49.7	2.10	0.12	−0.12	0.12	44.8	19.0	63.8	81.7	+96.2	+78
BI_2											
B—I'	49.7	2.10	0.16	−0.08	0.12	44.8	19.0	127.6	131.0	+57.9	+54
BI_3											
B—I'	49.7	2.03	0.18	−0.06	0.12	46.3	19.6	197.7	194.1	+13.4	+17
$B_3N_3H_6$											
B—H	84.1	(1.12)	0.19	0.01	0.09	77.8	26.7	313.5			
N—H	63.7	(1.02)	−0.20	0.01	0.11	58.9	35.8	284.1			
B—N'	51.8	1.44	0.19	−0.20	0.19	45.5	43.8	267.9			
B—N''	67.7	1.44	0.19	−0.20	0.19	59.4	43.8	309.6			
								1175.1	1177.4	−120.0	−122
$B_3N_3H_3Cl_3$											
B—Cl'''	73.3	(1.75)	0.27	−0.23	0.25	56.9	47.4	312.9			
N—H	61.0	(1.02)	−0.14	0.09	0.11	56.5	35.8	376.9			
B—N''	67.7	1.44	0.27	−0.14	0.20	58.7	46.1	628.8			
								1218.6	1223.9	−232.5	−237
$B_3O_3HCl_2$											
B—H	78.8	(1.10)	0.32	0.13	0.09	74.3	27.2	101.5			
B—O''	67.6	(1.38)	0.32	−0.24	0.28	54.3	67.4	730.2			
B—Cl'''	73.3	(1.75)	0.32	−0.19	0.25	56.8	47.4	208.4			
								1040.1	1031.6	−347.5	−339
$B_3O_3H_2Cl$											
B—H	80.6	(1.10)	0.28	0.09	0.09	76.1	27.2	206.6			
B—O''	67.6	(1.38)	0.28	−0.27	0.27	55.1	65.0	720.6			
B—Cl'''	73.3	(1.75)	0.28	−0.22	0.25	56.8	47.4	104.2			
								1031.4	1029.6	−315.8	−314

(continued

ABLE A—*continued*

Compound Bond	E(AB)	R_o	δ_A	δ_B	t_i	E_c	E_i	E_{at} (kpm) Calc.	Exp.	$\Delta Hf°$ (kpm) Calc.	Exp.
$_3O_3H_2F$											
B—H	79.7	(1.10)	0.30	0.11	0.09	75.2	27.2	204.8			
B—O″	67.6	(1.38)	0.30	− 0.26	0.28	54.3	67.3	729.6			
B—F‴	82.9	(1.30)	0.30	− 0.35	0.33	65.4	84.3	149.7			
								1084.1	1088.4	− 378.7	− 383
$_3O_3HF_2$											
B—H	77.0	(1.10)	0.36	0.17	0.10	71.8	30.2	102.0			
B—O″	67.6	(1.38)	0.36	− 0.21	0.29	53.6	69.7	739.8			
B—F‴	82.9	(1.30)	0.36	− 0.31	0.33	65.4	84.3	299.4			
								1141.2	1148.2	− 469.0	− 476
$_3O_3(OH)_3$											
O—H	55.3	(0.96)	− 0.23	0.14	0.19	48.5	65.7	342.6			
B—O″	67.6	(1.36)	0.33	− 0.23	0.28	55.1	68.4	1111.5			
								1454.1	1463.2	− 536.7	− 545.3
$_3O_3F_3$											
B—O″	67.6	(1.36)	0.42	− 0.16	0.29	54.4	70.8	751.2			
B—F‴	82.9	(1.30)	0.42	− 0.26	0.34	64.5	86.8	453.9			
								1205.1	1205.8	− 566.1	− 566.8
$_3O_3Cl_3$											
B—O″	67.6	(1.36)	0.36	− 0.21	0.29	54.4	70.8	751.2			
B—Cl‴	73.3	(1.75)	0.36	− 0.15	0.26	56.1	49.3	316.2			
								1067.4	1061.0	− 396.8	− 390.2
$_3H_3CO$											
B—H	82.8	1.19	0.22	0.04	0.09	72.2	25.1	291.9			
B—C	74.8	1.54	0.22	− 0.02	0.12	68.0	25.9	93.9			
C≡O‴	91.0	1.13	− 0.02	− 0.32	0.15	180.5	78.0	258.5			
								644.3	548.3		− 26.6
$_3H_3N(CH_3)_3$											
B—H	84.1	(1.14)	0.19	0.01	0.09	76.5	26.2	308.1			
B—N′	51.8	(1.62)	0.19	− 0.20	0.20	39.9	41.0	80.9			
C—N′	57.2	1.53	− 0.05	− 0.20	0.08	52.0	17.4	208.2			
C—H	(98.7)							888.3			
								1485.5	1406.9		− 20.3
$_3H_3PF_3$											
B—H	77.8	(1.10)	0.34	0.15	0.09	73.3	27.2	301.5			
B—P	59.2	(1.92)	0.34	0.18	0.08	54.5	13.8	68.3			
P—F‴	71.7	(1.59)	0.18	− 0.33	0.25	61.2	52.1	339.9			
								709.7	626.8		− 204.1
$_3H_3S(CH_3)_2$											
B—H	84.4	(1.14)	0.18	0.01	0.09	76.8	26.2	309.0			
B—S′	61.3	(1.85)	0.18	− 0.13	0.16	51.8	28.7	80.5			

(*continued*)

TABLE A—*continued*

Compound Bond	$E(AB)$	R_o	δ_A	δ_B	t_i	E_c	E_i	E_{at} (kpm) Calc.	Exp.	$\Delta Hf°$ (kpm) Calc.	Ex
C—S	67.7	(1.82)	− 0.05	− 0.13	0.04	64.6	7.3	143.8			
C—H	(98.7)							592.2			
								1125.5	1023.2		− 10
$BH_3S(C_2H_5)_2$											
B—H	84.1	(1.14)	0.19	0.01	0.09	76.6	26.2	308.4			
B—S′	61.4	(1.85)	0.19	− 0.13	0.16	51.8	28.7	80.5			
C—S	67.7	(1.82)	− 0.05	− 0.13	0.04	64.6	7.3	143.8			
C—C	83.2	(1.54)	− 0.05	− 0.05	0	83.2	0	166.4			
C—H	(98.7)							987.0			
								1686.1	1594.1		− 30
$(CH_3)_3BNH_3$											
C—H	(98.7)							888.3			
B—C	75.4	(1.56)	0.19	− 0.05	0.12	67.7	25.3	279.0			
B—N	51.8	(1.62)	0.19	− 0.20	0.20	40.0	41.0	81.0			
N—H	63.7	(1.02)	− 0.20	0.01	0.11	58.9	35.8	284.1			
								1531.4	1440.7		
$(CH_3)_3BNH_2CH_3$											
C—H	(98.7)							1184.4			
B—C	75.4	(1.56)	0.19	− 0.05	0.12	67.7	25.3	279.0			
B—N′	51.8	(1.62)	0.19	− 0.20	0.20	39.9	41.0	80.9			
C—N′	57.2	(1.47)	− 0.05	− 0.20	0.08	54.1	18.1	72.2			
N—H	63.7	(1.02)	− 0.20	0.01	0.11	58.9	35.8	189.4			
								1805.9	1714.7		− 52
$(CH_3)_3BN(CH_3)_3$											
C—H	(98.7)							1776.6			
B—C	75.4	(1.56)	0.19	− 0.04	0.12	67.7	25.3	279.0			
B—N	51.8	(1.62)	0.19	− 0.20	0.20	39.9	41.0	80.9			
C—N	57.2	(1.47)	− 0.04	− 0.20	0.08	54.1	18.1	216.6			
								2353.1	2265.7		− 52
$(CH_3)_3BNH(C_2H_5)_2$											
C—H	(98.7)							1875.3			
B—C	75.4	(1.56)	0.19	− 0.04	0.12	67.7	25.3	279.0			
B—N′	51.8	(1.62)	0.19	− 0.20	0.20	39.9	41.0	80.9			
C—N	57.2	(1.47)	− 0.04	− 0.20	0.08	54.1	18.1	144.4			
C—C	83.2	(1.54)	− 0.04	− 0.04	0	83.2	0	166.4			
N—H	63.5	(1.02)	− 0.20	0.01	0.11	58.8	35.8	94.6			
								2640.6	2551.5		− 62
$(CH_3)_3BP(CH_3)_3$											
C—H	(98.7)							1776.6			
B—C	75.4	(1.56)	0.18	− 0.05	0.12	67.7	25.3	279.0			
B—P′	59.2	(1.92)	0.18	0.04	0.07	55.1	12.1	67.2			

(*continued*)

ABLE A—*continued*

Compound Bond	$E(AB)$	R_o	δ_A	δ_B	t_i	E_c	E_i	E_{at} (kpm) Calc.	E_{at} (kpm) Exp.	$\Delta Hf°$ (kpm) Calc.	$\Delta Hf°$ (kpm) Exp.
P—C	65.2	(1.87)	0.04	−0.05	0.05	62.0	8.9	212.7			
								2335.5	2243.6		−68.3
$F_3O(CH_3)_2$											
C—H	(98.7)							592.2			
B—F‴	82.8	1.43	0.32	−0.34	0.33	59.4	76.6	408.0			
B—O′	48.2	1.50	0.32	−0.24	0.28	35.7	62.0	97.7			
C—O′	53.2	1.45	0.07	−0.24	0.15	46.5	34.3	161.6			
								1259.5	1234.5		
$F_3O(C_2H_5)_2$											
C—H	(98.7)							987.0			
B—F‴	82.8	(1.43)	0.28	−0.37	0.33	59.4	76.6	408.0			
B—O′	48.2	(1.50)	0.28	−0.27	0.28	35.7	62.0	97.7			
C—O′	53.2	(1.45)	0.03	−0.27	0.15	46.5	34.3	161.6			
C—C	83.2	(1.54)	0.03	0.03	0	80.7	0	161.4			
								1815.7	1800.8	−358.7	−343.8
$F_3N(CH_3)_3$											
C—H	(98.7)							888.3			
B—F‴	82.8	1.39	0.28	−0.37	0.32	62.0	76.4	415.2			
B—N′	51.8	1.58	0.28	−0.13	0.20	40.9	42.1	83.0			
C—N′	57.2	1.50	0.03	−0.13	0.08	53.0	17.7	212.1			
								1598.6	1591.4	−311.6	−304.4
$F_3P(CH_3)_3$											
C—H	(98.7)							888.3			
B—F‴	82.8	(1.43)	0.26	−0.38	0.32	60.3	74.3	403.8			
B—P	59.2	(1.92)	0.26	0.11	0.07	55.1	12.1	67.2			
C—P	65.2	(1.87)	0.02	0.11	0.05	61.9	8.9	212.4			
								1571.7	1561.8	−317.9	−312.6
AlH											
Al—H	68.6	1.65	0.19	−0.19	(0)	65.7	0	65.7	68.1	+64.4	+62.0
AlB_3H_{12}											
B—H	84.5	1.21	0.11	−0.06	0.08	73.2	21.9	570.6			
B—H	84.5	1.28	0.11	−0.06	0.08	69.2	20.7	269.7	(6 half-bonds)		
Al—H	68.6	(1.76)	0.36	−0.06	0.21	48.7	39.6	264.9	(6 half-bonds)		
								1105.2	1103.7	+1.5	+3.
AlO											
Al—O″	54.8	1.62	0.38	−0.38	0.38	41.5	77.8	119.3	115.8	+18.3	+21.8
AlF											
Al—F‴	67.2	1.65	0.43	−0.43	0.43	45.7	86.5	132.2	158.2	−35.3	−61.3
AlF_2											
Al—F‴	67.2	1.64	0.62	−0.31	0.47	42.8	95.1	275.8	272.8	−160.0	−157.0
AlF_3											
Al—F‴	67.2	1.63	0.72	−0.24	0.48	42.2	97.8	420.0	421.2	−285.3	−286.5

(continued)

TABLE A—*continued*

Compound Bond	E(AB)	R_o	δ_A	δ_B	t_i	E_c	E_i	E_{at} (kpm) Calc.	E_{at} (kpm) Exp.	$\Delta Hf°$ (kpm) Calc.	$\Delta Hf°$ (kpm) Exp.
AlCl₃											
Al—Cl‴	59.5	2.06	0.58	−0.19	0.39	39.6	62.9	307.5	305.0	−142.2	−139.
AlBr₃											
Al—Br‴	51.1	2.27	0.50	−0.17	0.33	36.2	48.3	253.5	256.4	−95.4	−98.
AlI₃											
Al—I‴	44.5	2.44	0.36	−0.12	0.24	35.9	32.7	205.8	203.5	−51.3	−49
GaH											
Ga—H	69.6	(1.58)	0.03	−0.03	(0)	69.6	0	69.6	65.6	+48.7	+52
GaCl₃											
Ga—Cl‴	60.3	2.09	0.31	−0.10	0.21	51.3	33.4	254.1	260.5	−100.6	−107
InH											
In—H	60.5	1.84	0.09	−0.09	(0)	57.5	0	57.5	45.8	+39.8	+51
InCl₃											
In—Cl‴	52.4	2.46	0.41	−0.14	0.27	37.6	36.4	222.0	221.3	−90.1	−89
InBr₃											
In—Br‴	45.1	2.58	0.34	−0.11	0.23	34.6	29.6	192.6	192.7	−67.3	−67
InI₃											
In—I‴	39.2	(2.80)	0.21	−0.07	0.14	33.2	16.6	149.4	150.5	−27.7	−28
TlF											
Tl—F′	26.0	2.08	0.49	−0.49	0.49	14.0	78.2	92.2	106.1	−29.7	−43
TlCl											
Tl—Cl′	37.4	2.48	0.40	−0.40	0.40	22.3	53.5	75.8	88.9	−3.1	−16
TlBr											
Tl—Br′	32.1	2.62	0.36	−0.36	0.36	20.5	45.6	66.1	79.3	+13.2	−9
TlI											
Tl—I′	28.0	2.81	0.28	−0.28	0.28	20.2	33.1	53.3	67.4	+15.8	+1
C₂											
C≡C	83.2	1.31	0	0	0	150.6	0	150.6	150.		
HCN											
H—C	89.0	1.07	0.09	0.03	0.03	87.9	9.3	97.2			
C≡N‴	88.9	1.16	0.03	−0.13	0.08	159.7	34.3	194.0			
								291.2	304.1	+45.2	+32
(CN)₂											
C—C	83.2	1.38	0.08	0.08	0	85.4	0	85.4			
C≡N‴	88.9	1.16	0.08	−0.08	0.08	159.7	34.3	388.0			
								473.4			
CNF											
C—F′	56.2	1.26	0.20	−0.23	0.22	51.5	58.0	109.5			
C≡N‴	88.9	1.16	0.20	0.03	0.09	157.9	38.6	196.5			
								306.0	305.	−2.8	−2.
CNCl											
C—Cl′	69.6	1.63	0.15	−0.12	0.13	65.4	26.5	91.9			

(continued

ABLE A—*continued*

ompound Bond	$E(AB)$	R_o	δ_A	δ_B	t_i	E_c	E_i	E_{at} (kpm) Calc.	Exp.	ΔHf° (kpm) Calc.	Exp.
C≡N‴	88.9	1.16	0.15	−0.03	0.09	157.9	38.6	196.5			
								288.4	280.4	+25.0	+33.0
NBr											
C—Br′	62.0	1.79	0.12	−0.06	0.09	60.2	16.7	76.9			
C≡N‴	88.9	1.16	0.12	−0.05	0.08	159.7	34.3	194.0			
								270.9	266.5	+40.1	+44.5
NI											
C—I′	54.8	2.00	0.06	0.05	0	54.7	0	54.7			
C≡N‴	88.9	1.16	0.06	−0.11	0.08	159.7	34.3	194.0			
								248.7	255.9	+61.1	+53.9
O											
C≡O‴	91.0	1.13	0.16	−0.16	0.16	178.4	83.2	261.6	257.3	−30.7	−26.4
O₂											
C=O″	74.5	1.16	0.22	−0.11	0.17	119.1	73.0	384.2	384.6	−93.7	−94.1
oly CO₂											
C—O′	53.2	(1.45)	0.22	−0.11	0.17	45.4	38.9	337.2		−46.7	
₃O₂											
C=C	83.2	1.28	0.13	0.13	0	126.0	0	252.2			
C=O″	74.5	1.19	0.13	−0.19	0.16	117.5	67.0	369.0			
								621.2	609.7	+11.9	+23.4
₂CO₃											
O—H	53.3	(0.95)	−0.18	0.20	0.19	47.3	66.4	227.4			
C—O′	53.2	(1.38)	0.14	−0.18	0.16	48.2	38.5	173.4			
C=O″	74.5	(1.25)	0.14	−0.18	0.16	111.9	63.7	175.6			
								576.4		−122.1	
OS											
C=O″	74.5	1.16	0.13	−0.19	0.16	120.6	68.7	189.3			
C=S‴	75.8	1.56	0.13	0.05	0.04	126.6	12.8	139.4			
								328.7	331.5	−31.2	−34.0
S											
C≡S‴	75.8	1.53	0.04	−0.04	0.04	152.4	15.4	167.8	181.9	+70.1	+56.
S₂											
C=S‴	75.8	1.56	0.05	−0.03	0.04	126.6	12.8	278.8	276.4	+25.7	+28.1
COF											
H—C	81.3	1.10	0.24	0.17	0.03	78.1	9.1	87.2			
C=O″	74.5	1.18	0.17	−0.15	0.16	118.5	67.5	186.0			
C—F″	75.8	1.35	0.17	−0.26	0.21	65.6	51.6	117.2			
								390.4	391.9	−88.5	−90.0
OF₂											
C=O″	74.5	1.17	0.31	−0.03	0.17	118.1	72.4				
								169.8			

(*continued*)

TABLE A—*continued*

Compound Bond	E(AB)	R_o	δ_A	δ_B	t_i	E_c	E_i	E_{at} (kpm) Calc.	E_{at} (kpm) Exp.	$\Delta Hf°$ (kpm) Calc.	$\Delta Hf°$ (kpm) Exp.
C—O‴	91.0	1.17	0.31	−0.03	0.17	95.1	51.1				
C—F″	75.8	1.31	0.31	−0.14	0.23	65.9	58.3	248.4			
								418.2	420.4	−149.5	−151.
COCl₂											
C=O″	74.5	1.17	0.22	−0.13	0.18	116.7	76.6	169.8			
C—O‴	91.0	1.17	0.22	−0.13	0.18	95.1	51.1				
C—Cl′	69.6	1.75	0.22	−0.05	0.14	60.2	26.6	173.6			
								343.4	341.4	−54.3	−52.
CF											
C—F″	75.8	1.27	0.22	−0.22	0.22	68.9	57.5	126.4	115	+63.8	+75.
CF₂											
C—F″	75.8	1.30	0.30	−0.15	0.23	66.4	58.7	250.2	250.	−41.1	−41.
CF₃											
C—F′	56.1	(1.32)	0.34	−0.11	0.23	48.4	57.8	106.2			
C—F″	75.8	(1.32)	0.34	−0.11	0.23	65.4	57.8	246.4			
								352.6	344.5	−124.6	−116.
CF₄											
C—F′	56.1	1.32	0.37	−0.09	0.23	48.4	57.8	212.4			
C—F″	75.8	1.32	0.37	−0.09	0.23	65.4	57.8	246.4			
								458.8	465.	−211.7	−219.
C₂F₄											
C=C	83.2	1.31	0.30	0.30	0	146.7	0	146.7			
C—F′	56.1	1.31	0.30	−0.15	0.23	48.8	58.3	428.4			
								575.1	573.7	−156.9	−155.
C₂F₆											
C—C	83.2	1.54	0.34	0.34	0	54.9	0	54.9			
C—F′	56.1	1.32	0.34	−0.11	0.23	48.4	57.8	637.2			
								692.1	696.	−236.1	−240.
CCl											
C—Cl′	69.6	1.76	0.13	−0.13	0.13	60.6	24.5	85.1	68.4	+115.3	+132.0
CCl₄											
C—Cl′	69.6	1.77	0.23	−0.06	0.15	58.8	28.1	347.6	312.3	−59.9	−24.0
CCl₃F											
C—Cl′	69.6	1.76	0.26	−0.02	0.14	59.9	26.4	258.9			
C—F″	75.8	1.33	0.26	−0.18	0.22	65.8	54.9	120.7			
								379.6	345.5	−102.1	−68.0
CCl₂F₂											
C—Cl′	69.6	1.77	0.29	0.01	0	69.2	0	138.4			
C—F″	75.8	1.33	0.29	−0.15	0.22	65.8	54.9	241.4			
								379.8	382.3	−112.5	−115.0

(*continued*)

TABLE A—*continued*

Compound Bond	$E(AB)$	R_o	δ_A	δ_B	t_i	E_c	E_i	E_a (kpm) Calc.	E_a (kpm) Exp.	ΔHf° (kpm) Calc.	ΔHf° (kpm) Exp.
CClF₃											
C—Cl′	69.6	1.75	0.33	0.04	0	70.0	0	70.0			
C—F′	56.1	1.33	0.33	− 0.12	0.23	48.1	57.4	105.5			
C—F″	75.8	1.33	0.33	− 0.12	0.23	64.9	57.4	244.6			
								420.1	423.1	− 163.0	− 166.0
Si₂											
Si≡Si	53.4	2.25	0	0	0	83.2	0	83.2	75.8	+ 134.6	+ 142.
SiH											
Si—H	74.7	1.52	0.09	− 0.09	(0)	73.2	0	73.2	74.7	+ 87.8	+ 86.3
SiH₄											
Si—H	74.7	1.46	0.16	− 0.04	(0)	76.1	0	304.4	309.5	+ 12.9	+ 7.3
Si₂H₆											
Si—H	74.7	1.48	0.15	− 0.05	(0)	75.2	0	451.2			
Si—Si	53.4	2.32	0.15	− 0.15	(0)	53.8	0	53.8			
								505.0	511.2	+ 25.4	+ 19.2
Si₃H₈											
Si—H	74.7	1.48	0.14	− 0.05	(0)	75.2	0	601.6			
Si—Si	53.4	2.32	0.14	0.14	0	53.8	0	107.6			
								709.2	714.6	+ 34.3	+ 28.9
SiC											
Si—C	66.7	(1.65)	0.18	− 0.18	0.18	64.4	36.2	100.6	103.2		
SiN											
Si—N″	59.9	1.57	0.21	− 0.21	0.21	57.6	44.4	102.0	105.6	+ 119.9	+ 116.3
SiO											
Si=O‴	72.8	1.51	0.29	− 0.29	0.29	97.1	95.7	192.8	192.3	− 24.3	− 23.8
SiS											
Si≡S‴	60.8	1.93	0.17	− 0.17	0.17	102.3	51.8	154.1	153.6	+ 21.4	+ 16.9
SiO₂(g)											
Si—O‴	72.8	1.54	0.40	− 0.20	0.30	62.5	64.7	127.2			
Si=O″	59.7	1.54	0.40	− 0.20	0.30	76.9	97.0	173.9			
								301.1	304.3	− 73.0	− 76.2
SiO₂(c)											
Si—O″	59.7	1.61	0.40	− 0.20	0.30	49.1	61.8	443.6	445.8	− 215.5	− 217.7
SiF											
Si—F⁗	79.6	1.60	0.34	− 0.34	0.34	61.7	70.6	132.3	132.	− 4.5	− 4.
SiF₂											
Si—F⁗	79.6	1.49	0.48	− 0.24	0.36	64.3	80.2	289.0	289.	− 142.0	− 142.
SiF₃											
Si—F⁗	79.6	1.54	0.55	− 0.18	0.37	61.2	79.8	423.0	433.1	− 257.4	− 267.5
SiF₄											
Si—F⁗	79.6	1.54	0.61	− 0.15	0.38	60.2	81.9	568.4	570.5	− 383.9	− 386.0

(continued)

TABLE A—*continued*

Compound Bond	E(AB)	R_o	δ_A	δ_B	t_i	E_c	E_i	E_{at} (kpm) Calc.	E_{at} (kpm) Exp.	ΔHf° (kpm) Calc.	ΔHf° (kpm) Exp.
SiH₃F											
Si—H	73.5	1.47	0.26	0.05	0.10	67.1	22.6	269.1			
Si—F″	60.8	1.59	0.26	− 0.40	0.33	48.2	68.9	117.1			
								386.2	· 389.1	− 102.1	− 105.0
SiH₂F₂											
Si—H	68.8	1.47	0.37	0.15	0.11	62.1	24.8	173.8			
Si—F‴′	79.6	1.58	0.37	− 0.33	0.35	61.6	73.5	270.2			
								444.0	444.9	− 193.1	− 194.0
SiHF₃											
Si—H	64.7	1.46	0.48	0.25	0.12	58.1	27.3	85.4			
Si—F‴′	79.6	1.56	0.48	− 0.24	0.36	61.4	76.6	414.3			
								499.7	500.7	− 282.0	− 283.0
SiCl											
Si—Cl‴	64.8	2.00	0.26	− 0.26	0.26	51.8	43.2	95.0	92.6	+ 43.0	+ 45.4
SiCl₂											
Si—Cl‴	64.8	2.00	0.36	− 0.18	0.27	51.1	44.8	191.8	206.7	− 24.7	− 39.6
SiCl₄											
Si—Cl‴	64.8	2.02	0.43	− 0.11	0.27	50.6	44.4	380.0	382.3	− 154.7	− 157.0
SiH₃Cl											
Si—H	73.9	1.48	0.23	0.02	0.10	67.0	22.4	268.2			
Si—Cl′	55.8	2.05	0.23	− 0.28	0.25	44.1	40.5	84.6			
								352.8	342.3	− 58.5	− 48.0
SiH₂Cl₂											
Si—H	71.6	1.48	0.29	0.08	0.11	64.2	24.7	177.8			
Si—Cl′	55.8	2.05	0.29	− 0.23	0.26	43.5	42.1	171.2			
								349.0	346.3	− 77.7	− 75.0
SiHCl₃											
Si—H	68.8	(1.49)	0.37	0.15	0.11	61.2	24.5	85.7			
SiCl″	60.4	2.02	0.37	− 0.17	0.27	47.1	44.4	274.5			
								360.2	360.3	− 111.9	− 112.0
SiBr₄											
Si—Br‴	55.6	2.15	0.36	− 0.09	0.23	46.0	35.5	326.0	315.0	− 110.3	− 99.3
GeH₄											
Ge—H	71.7	1.54	− 0.01	0.00	0	71.7	0	286.8	276.7	+ 11.6	+ 21.7
GeC											
Ge≡C	64.0	(1.85)	0.03	− 0.03	0.03	100.2	8.1	108.3	110.3	+ 153.0	+ 151.
GeO											
Ge≡O‴	69.9	1.65	0.19	− 0.19	0.19	99 9	57.3	157.2	160.6	− 7.6	− 11.0
GeF											
Ge—F‴	70.2	1.68	0.24	− 0.24	0.24	61.3	47.4	108.7	116.9	+ 0.2	− 8.0
GeF₄											

(continued)

TABLE A—*continued*

Compound Bond	$E(AB)$	R_o	δ_A	δ_B	t_i	E_c	E_i	E_{at} (kpm) Calc.	E_{at} (kpm) Exp.	ΔHf° (kpm) Calc.	ΔHf° (kpm) Exp.
Ge—F'''	70.2	1.68	0.42	−0.10	0.26	59.8	51.4	444.8	450.0	−279.2	−284.4
GeCl											
Ge—Cl'''	62.2	(2.08)	0.16	−0.16	0.16	55.5	25.6	81.1	82.1	+38.0	+37.
GeCl₄											
Ge—Cl'''	62.2	2.09	0.26	−0.07	0.16	55.2	25.4	322.4	324.9	−116.0	−118.5
GeBr											
Ge—Br'''	53.4	2.30	0.11	−0.11	0.11	48.8	15.9	64.7	60.4	+52.0	+56.3
GeBr₂											
Ge—Br'''	53.4	(2.30)	0.15	−0.08	0.12	48.2	17.3	131.0	158.4	+12.4	−15.0
GeBr₄											
Ge—Br'''	53.4	2.30	0.18	−0.05	0.11	48.8	15.9	258.8	268.5	−62.0	−71.7
GeI₄											
Ge—I'''	46.5	2.49	0.05	−0.01	0.03	46.2	4.0	200.8	205.6	−8.8	−13.6
SnH											
Sn—H	62.3	1.79	0.06	−0.06	(0)	59.9	0	59.9	74.	+64.4	+50.3
SnH₄											
Sn—H	62.3	1.70	0.10	−0.02	(0)	63.0	0	252.0	241.7	+28.6	+38.9
SnCl₄											
Sn—Cl'''	54.1	2.31	0.38	−0.10	0.24	42.5	34.5	308.0	307.1	−119.4	−118.5
SnBr₄											
Sn—Br'''	46.5	2.44	0.30	−0.08	0.19	39.2	25.9	260.4	254.2	−81.4	−75.2
PbF											
Pb—F'''	45.5	(2.11)	0.41	−0.41	0.41	27.7	64.5	92.2	74.7	−26.5	−9.0
PbF₂											
Pb—F'''	45.5	2.11	0.59	−0.29	0.44	26.3	69.3	191.2	188.6	−106.6	−104.0
PbF₄											
Pb—F'	27.5	2.08	0.54	−0.14	0.34	19.0	54.3	146.6			
Pb—F''	37.7	2.08	0.54	−0.14	0.34	26.1	54.3	160.8			
								307.4	308.4	−185.0	−186.0
PbCl											
Pb—Cl'''	40.1	(2.46)	0.32	−0.32	0.32	27.3	43.2	70.5	72.4	+5.4	+3.5
PbCl₂											
Pb—Cl'''	40.1	2.46	0.46	−0.23	0.35	26.1	47.2	146.6	145.6	−41.6	−40.6
PbCl₄											
Pb—Cl'	34.5	2.43	0.38	−0.09	0.24	26.6	32.8	237.6	238.2	−74.4	−75.0
PbBr											
Pb—Br'''	34.5	(2.60)	0.28	−0.28	0.28	24.9	35.7	60.6	60.2	+12.9	+13.3
PbBr₂											
Pb—Br'''	34.5	2.60	0.40	−0.20	0.30	24.2	38.3	125.0	124.4	−24.8	−24.2
PbBr₄											
Pb—Br'	30.8	2.58	0.30	−0.08	0.19	25.3	24.4	198.8	197.6	−45.2	−44.0
PbI											
Pb—I'''	30.0	(2.79)	0.20	−0.20	0.20	24.1	23.8	47.9	46.6	+24.4	+25.7

(*continued*)

TABLE A—*continued*

Compound Bond	E(AB)	R_o	δ_A	δ_B	t_i	E	E_i	E_{at} (kpm) Calc.	E_{at} (kpm) Exp.	$\Delta Hf°$ (kpm) Calc.	$\Delta Hf°$ (kpm) Exp.
PbI_2											
Pb—I'''	30.0	2.79	0.28	− 0.14	0.21	23.8	25.0	97.6	97.6	+ 0.2	+ 0.2
PbI_4											
Pb—I'	27.2	2.77	0.16	− 0.04	0.10	24.8	12.0	147.2	149.2	+ 1.6	− 0.4
N_2											
N'''≡N'''	94.8	1.10	0	0	0	225.8	0	225.8	226.0	+ 0.2	0
NH											
N—H	60.3	1.04	− 0.11	0.11	0.11	54.7	35.1	89.8	85.9	+ 75.3	+ 79.2
NH_2											
N—H	61.6	1.03	− 0.15	0.07	0.11	56.4	35.5	183.8	176.9	+ 33.4	+ 40.3
NH_3											
N—H	62.3	1.02	− 0.16	0.05	0.11	57.6	35.8	280.2	280.3	− 10.9	− 11.0
N_2H_4											
N—H	61.7	1.02	− 0.15	0.07	0.11	57.1	35.8	371.6			
N—N	39.2	1.43	− 0.15	− 0.15	0	40.0	0	40.0			
								411.6	411.6	+ 22.8	+ 22.8
HN_3											
N—H	58.3	1.02	0.17	− 0.06	0.12	53.3	39.1	92.4			
HN''=N''	67.0	1.24	− 0.06	− 0.06	0	120.0	0	120.0			
N''—N'''	79.7	1.13	− 0.06	− 0.06	0	104.4	0	104.4			
								316.8	320.8	+ 74.3	+ 70.3
N_2H_5OH											
N—N	39.2	(1.45)	− 0.13	− 0.13	0	40.0	0	40.0			
N—H	61.0	(1.02)	− 0.13	0.09	0.11	56.5	35.8	461.5			
N—O'	36.5	(1.44)	− 0.13	− 0.28	0.07	34.4	16.1	50.5			
O—H	56.8	(0.96)	− 0.28	0.09	0.18	50.5	62.2	112.7			
								664.7	647.2	− 66.5	− 49.0
N_2O											
N'''—N'''	90.1	1.13	0.05	0.05	0	118.0	0	118.0			
N''=O''	66.8	1.19	0.05	− 0.10	0.08	113.1	33.5	146.6			
								264.6	266.1	+ 21.0	+ 19.5
NO											
N''=O''	66.8	1.15	0.08	− 0.08	0.08	117.0	34.6	151.6	151.0	+ 21.0	+ 21.6
NO_2											
N'—O''	51.2	1.19	0.11	− 0.05	0.08	57.8	22.3	80.1			
N''=O''	66.8	1.19	0.11	− 0.05	0.08	113.1	33.5	146.6			
								226.7	224.2	+ 5.5	+ 3.0
HNO_2											
H—O	51.2	0.96	0.26	− 0.14	0.20	44.4	69.2	113.6			
N'—O'	36.5	1.44	0.02	− 0.14	0.08	34.0	18.4	52.4			
N''=O''	66.8	1.20	0.02	− 0.14	0.08	112.1	33.2	145.3			
								311.3	303.2	− 27.0	− 18.9

(*continued*)

TABLE A—*continued*

Compound Bond	$E(AB)$	R_o	δ_A	δ_B	t_i	E_c	E_i	E_{at} (kpm) Calc.	E_{at} (kpm) Exp.	$\Delta Hf°$ (kpm) Calc.	$\Delta Hf°$ (kpm) Exp.
HNO₃											
H—O	50.2	0.96	0.29	−0.11	0.20	43.5	69.2	112.7			
N′—O′	36.5	1.44	0.04	−0.11	0.08	34.0	18.4	52.4			
N′—O′	36.5	1.20	0.04	−0.11	0.08	40.9	22.1	63.0			
N″=O″	66.8	1.20	0.04	−0.11	0.08	112.2	33.2	145.4			
								373.5	376.2	− 29.6	− 32.3
H₃NO₄											
H—O	54.6	(0.96)	0.22	−0.16	0.19	47.9	65.7	340.8			
N′—O′	36.5	(1.44)	− 0.01	−0.16	0.08	34.0	18.4	209.6			
								550.4		− 42.8	
NOF											
N′—F′	38.5	1.52	0.14	− 0.13	0.13	31.9	28.4	60.3			
N″=O″	66.8	1.13	0.14	− 0.02	0.08	119.1	35.3	154.4			
								214.7	207.2	− 23.2	− 15.7
NS											
S″=N″	64.6	1.50	0.04	− 0.04	0.04	110.4	13.3	123.7	115?	+ 55.9	+ 65?
NF											
N′—F′	38.5	1.37	0.14	− 0.14	0.14	35.0	33.9	68.9	73.3	+ 63.0	+ 58.6
NF₂											
N′—F′	38.5	1.37	0.18	− 0.09	0.14	35.0	33.9	137.8	140.7	+ 13.0	+ 10.1
NF₃											
N′—F′	38.5	1.37	0.20	− 0.07	0.14	35.0	33.9	206.7	200	− 37.0	− 30.
N₂F₄											
N′—N′	39.2	1.47	0.18	0.18	0	32.4	0	32.4			
N′—F′	38.5	1.37	0.18	− 0.09	0.14	35.0	33.9	275.6			
								308.0	303.3	− 6.4	− 1.7
NCl₃											
N′—Cl′	47.8	(1.73)	0.07	− 0.02	0.05	45.4	9.6	165.0		+ 35.3	
NBr₃											
N′—Br′	42.6	(1.87)	0.07	− 0.02	0.05	40.7	8.9	148.8		+ 44.3	
NI₃											
N′=I′	37.6	(2.06)	− 0.11	0.04	0.07	35.2	11.3	139.5		+ 50.0	
P₂											
P‴≡P‴	60.7	1.89	0	0	0	125.1	0	125.1	125.1	+ 34.5	+ 34.5
P₄											
P′—P′	51.1	2.21	0	0	0	50.9	0	305.4	305.1	+ 13.8	+ 14.1
PH											
P—H	73.1	(1.42)	0.01	− 0.01	0.01	72.4	2.3	74.7	72.3	+ 57.2	+ 55.0
PH₂											
P—H	73.1	(1.42)	0.02	− 0.01	0.01	72.4	2.3	149.4	153.5	+ 30.0	+ 25.9
PH₃											
P—H	73.1	1.42	0.02	− 0.01	0.02	71.6	4.7	228.9	230.2	+ 2.6	+ 1.3

(*continued*)

TABLE A—*continued*

Compound Bond	$E(AB)$	R_o	δ_A	δ_B	t_i	E_c	E_i	E_{at} (kpm) Calc.	Exp.	$\Delta Hf°$ (kpm) Calc.	Exp.
P_2H_4											
P—H	73.1	(1.42)	0.02	−0.01	0.02	71.6	4.7	305.2			
P—P	51.1	(2.20)	0.02	0.02	0	51.1	0	51.1			
								356.3	353.8	+2.5	+5.◖
PN											
P″=N‴	75.9	1.49	0.13	−0.13	0.13	117.3	43.5	160.8	155.4	+27.4	+32.◖
$(PN(CH_3)_2)_3$											
C—H	91.8	1.09	−0.03	0.03	0.03	89.0	9.1	1765.8			
C—P′	65.3	(1.85)	−0.03	0.06	0.05	62.7	9.0	430.2			
P′—N′	44.8	(1.65)	0.06	−0.19	0.12	44.0	24.1	204.3			
P′—N″	58.6	(1.65)	0.06	−0.19	0.12	57.5	24.1	244.8			
								2645.1	2637.3	−114.9	−107.◖
$(PN(C_6H_5)_2)_4$											
C—C	83.2	(1.39)	−0.02	−0.02	0	122.6	0	5884.8			
C—H	91.3	(1.08)	−0.02	0.04	0.03	89.4	9.2	3944.0			
C—P′	65.3	(1.81)	−0.02	0.07	0.05	64.1	9.2	586.4			
P′—N′	44.8	(1.65)	0.07	−0.18	0.12	44.0	24.1	272.4			
P′—N″	58.6	(1.65)	0.07	−0.18	0.12	57.5	24.1	326.4			
								11014.0	11000.5	+45.2	+58.7
$(PNCl_2)_3$											
P′—Cl′	54.7	1.97	0.24	−0.11	0.18	47.6	40.3	467.4			
P′—N′	44.8	1.65	0.24	−0.02	0.13	43.5	26.2	209.1			
P′—N″	58.6	1.65	0.24	−0.02	0.13	56.9	26.2	249.3			
								925.8	915.1	−186.6	−175.◖
$(PNCl_2)_4$											
P′—Cl′	54.7	1.97	0.24	−0.11	0.18	47.6	40.3	623.2			
P′—N′	44.8	1.65	0.24	−0.02	0.13	43.5	26.2	278.8			
P′—N″	58.6	1.65	0.24	−0.02	0.13	56.9	26.2	332.4			
								1234.4	1221.7	−248.8	−236.◖
PO											
P′—O‴	77.8	1.45	0.21	−0.21	0.21	70.7	48.1	118.8	120.4	+16.0	+14.◖
P_4O_6											
P′—O″	58.4	1.65	0.25	−0.17	0.21	50.9	42.3	559.2			
P′—O″	71.2	1.65	0.25	−0.17	0.21	62.5	42.3	628.8			
								1188.0	1187.5	−529.8	−529.◖
P_4O_{10}											
P′—O′	41.7	1.62	0.31	−0.12	0.22	36.6	45.1	490.2			
P′—O″	58.4	1.62	0.31	−0.12	0.22	51.2	45.1	577.8			
P‴—O‴	77.8	1.39	0.31	−0.12	0.22	79.5	52.5	528.0			
								1596.0	1588.7	−699.2	−691.◖

(*continued*)

TABLE A—*continued*

Compound Bond	$E(AB)$	R_o	δ_A	δ_B	t_i	E_c	E_i	E_{at} (kpm) Calc.	E_{at} (kpm) Exp.	$\Delta Hf°$ (kpm) Calc.	$\Delta Hf°$ (kpm) Exp.
PS											
P‴=S‴	64.7	1.92	0.09	− 0.09	0.09	98.5	23.4	121.9	123.5	+ 19.9	+ 18.3
PF											
P′—F‴	71.7	1.59	0.26	− 0.26	0.26	60.4	54.3	114.7	112.3	− 20.6	− 18.2
PF₂											
P′—F‴	71.7	(1.55)	0.36	− 0.18	0.27	61.1	57.8	237.8	222.2	− 124.8	− 109.2
PF₃											
P′—F‴	71.7	1.54	0.42	− 0.14	0.28	60.7	60.4	363.3	357.9	− 231.4	− 226.0
PF₅											
P′—F′	44.0	1.57	0.47	− 0.09	0.28	36.5	59.2	191.4			
P′—F‴	71.7	1.57	0.47	− 0.09	0.28	59.5	59.2	363.0			
								554.4	555.1	− 384.7	− 381.4
PCl											
P′—Cl′	54.5	2.04	0.18	− 0.18	0.18	45.8	29.3	75.1	73.5	+ 29.2	+ 30.8
PCl₃											
P′—Cl′	54.5	2.04	0.27	− 0.09	0.18	45.8	29.3	225.3	231.1	− 62.8	− 68.6
PCl₅											
P′—Cl′	54.5	2.19	0.32	− 0.06	0.19	42.2	28.8	71.0	(2 half-bonds)		
P′—Cl″	59.2	2.04	0.32	− 0.06	0.19	49.1	30.9	240.0			
								311.0	310.3	− 90.3	− 89.6
PBr											
P′—Br′	48.6	2.23	0.13	− 0.13	0.13	42.5	19.4	61.9	63.1	+ 40.0	+ 38.8
PBr₃											
P′—Br′	48.6	2.20	0.22	− 0.07	0.14	42.5	21.1	190.8	188.7	− 35.4	− 33.3
PI₃											
P′—I′	43.0	2.47	0.08	− 0.03	0.05	40.2	6.7	140.7		+ 11.0	
POF₃											
P′—O‴	71.3	1.45	0.43	− 0.03	0.23	68.9	52.7	121.6			
P′—F‴	71.7	1.52	0.43	− 0.13	0.28	61.5	61.2	368.1			
								489.7	481.0	− 298.2	− 289.5
POCl₃											
P′—O‴	71.3	1.45	0.32	− 0.12	0.22	69.8	50.4	120.2			
P′—Cl′	54.6	1.99	0.32	− 0.07	0.19	46.4	31.7	234.3			
								354.5	355.6	− 132.4	− 133.5
POBr₃											
P′—O″	58.4	1.45	0.25	− 0.17	0.21	57.9	48.1	106.0			
P′—Br′	48.7	2.06	0.25	− 0.03	0.14	45.5	22.6	204.3			
								310.3	312.2	− 95.3	− 97.2
PSCl₃											
P′—S″	56.4	1.85	0.26	0.07	0.09	59.4	16.2	75.6			
P′—Cl″	59.2	2.02	0.26	− 0.11	0.18	50.2	29.6	239.4			
								315.0	320.1	− 85.9	− 91.0

(continued)

TABLE A—*continued*

Compound Bond	E(AB)	R_0	δ_A	δ_B	t_i	E_c	E_i	E_{at} (kpm) Calc.	E_{at} (kpm) Exp.	ΔHf° (kpm) Calc.	ΔHf° (kpm) Exp.
AsH_3											
H—As	66.6	(1.51)	0.02	−0.06	0.04	64.0	8.8	218.4	212.7	+10.2	+15.9
As_4O_6											
As′—O″	53.8	1.79	0.19	−0.13	0.16	48.2	29.7	934.8	935.8	−288.0	−289.0
AsF_3											
As′—F‴	66.0	1.71	0.32	−0.11	0.21	58.0	40.8	296.4	349.0	−167.4	−220.0
AsF_5											
As′—F′	40.5	(1.71)	0.38	−0.08	0.23	34.6	44.7	158.6			
As′—F‴	66.0	(1.71)	0.38	−0.08	0.23	56.5	44.7	303.6			
								462.2	462.4	−295.4	−295.6
$AsCl_3$											
As′—Cl‴	58.5	2.16	0.20	−0.07	0.13	51.4	20.0	214.2	221.4	−54.6	−61.8
$AsBr_3$											
As′—Br‴	50.3	2.33	0.13	−0.04	0.09	45.8	12.8	175.8	183.5	−23.3	−31.
AsI_3											
As′—I‴	43.6	2.52	0	0	0	43.6	0	130.8		+18.0	
$SbCl_3$											
Sb—Cl‴	50.2	2.33	0.30	−0.10	0.20	40.9	28.5	208.2	225.0	−58.2	−75.0
$SbBr_3$											
Sb—Br‴	43.2	2.50	0.23	−0.08	0.15	37.0	19.9	170.7	189.3	−27.9	−46.5
SbH_3											
Sb—H	57.8	1.71	0.04	−0.01	0.03	55.7	5.8	184.5	184.3	+34.5	+34.7
$BiCl_3$											
Bi—Cl‴	48.9	2.48	0.33	−0.11	0.22	37.7	29.4	201.3	200.0	−64.5	−63.2
$BiBr_3$											
Bi—Br‴	41.9	2.63	0.26	−0.09	0.17	34.4	21.4	167.4	168.1	−37.8	−38.5
BiI_3											
Bi—I‴	36.2	2.84	0.13	−0.04	0.09	32.4	10.5	128.7	129.9	−2.7	−3.9
O_2											
O″=O″	66.7	1.21	0	0	0	119.1	0	119.1	119.2	0	0
O_3											
O″—O′	50.4	1.28	0	0	0	56.7	0	56.7			
O″—O′	50.4	1.28	0	0	0	85.1	0	85.1			
								141.8	143.5	+35.8	+34.1
OF											
O′—F′	35.9	1.42	0.05	−0.05	0.05	34.3	11.7	46.0	46.1	+32.5	+32.4
OF_2											
O′—F′	35.9	1.42	0.07	−0.04	0.06	34.0	14.0	96.0	89.8	+1.4	+7.6
S_2											
S″=S″	62.1	1.89	0	0	0	102.5	0	102.5	102.4	+30.7	+30.8
S_2O											
S″=S″	62.1	1.88	0.08	0.08	0	94.8	0	94.8			
S′—O‴	73.9	1.47	0.08	−0.16	0.12	77.8	27.1	104.9			
								199.7		−6.9	

(*continued*)

TABLE A—*continued*

Compound Bond	$E(AB)$	R_o	δ_A	δ_B	t_i	E_c	E_i	E_{at} (kpm) Calc.	Exp.	$\Delta Hf°$ (kpm) Calc.	Exp.
SO											
$S''{=}O''$	64.4	1.49	0.12	−0.12	0.12	100.4	40.1				
								122.0	124.7	+4.2	+1.5
$S'{-}O'''$	73.9	1.49	0.12	−0.12	0.12	76.8	26.7				
SO$_2$											
$S''{=}O''$	64.4	1.43	0.16	−0.08	0.12	104.6	41.8	146.4			
$S'{-}O'''$	73.9	1.43	0.16	−0.08	0.12	80.0	27.9	107.9			
								254.3	256.7	−68.5	−70.9
SO$_3$											
$S''{-}O''$	64.4	1.43	0.18	−0.06	0.12	69.7	27.9	195.2			
$S''{=}O''$	64.4	1.43	0.18	−0.06	0.12	104.6	41.8	146.4			
								341.6	340.0	−96.0	−94.6
H$_2$SO$_3$											
$H{-}O$	52.6	(0.97)	0.22	−0.17	0.20	45.1	68.5	227.2			
$S'{-}O'$	43.3	(1.53)	0.07	−0.17	0.12	43.8	26.0	139.6			
$S'{-}O'''$	73.9	1.42	0.07	−0.17	0.12	80.6	28.1	108.7			
								475.5		−125.9	
H$_2$SO$_4$											
$H{-}O$	51.6	0.97	0.25	−0.15	0.20	44.3	68.5	225.6			
$S'{-}O'$	43.3	1.53	0.09	−0.15	0.12	43.8	26.0	139.6			
$S'{-}O'''$	73.9	1.42	0.09	−0.15	0.12	80.6	28.1	217.4			
								582.6	586.2	−173.4	−177.0
S$_2$F$_2$											
$S'{-}S'$	55.0	(2.03)	0.18	0.18	0	46.2	0	46.2			
$S'{-}F'$	45.6	(1.60)	0.18	−0.18	0.18	40.9	37.3	156.4			
								202.6		−31.6	
SF$_2$											
$S'{-}F'$	45.6	1.59	0.24	−0.12	0.18	41.2	37.6	157.6		−43.2	
SF$_4$											
$S'{-}F'$	45.6	1.56	0.30	−0.07	0.19	41.4	40.4	327.2	327.4	−185.0	−185.2
SF$_6$											
$S'{-}F'$	45.6	1.56	0.33	−0.05	0.19	41.4	40.4	490.8	471.8	−310.8	−291.8
SF$_5$Cl											
$S'{-}F'$	45.6	(1.56)	0.29	−0.08	0.19	41.4	40.4	409.0			
$S'{-}Cl'$	56.6	(2.00)	0.29	0.09	0.10	51.7	16.6	68.3			
								477.3	440.7	−286.1	−250.5
S$_2$Cl$_2$											
$S'{-}S'$	55.0	2.05	0.09	0.09	0	50.8	0	50.8			
$S'{-}Cl'$	56.6	1.99	0.09	−0.09	0.09	52.5	15.0	135.0			
								185.8	195.8	+5.6	−4.4

(continued)

TABLE A—*continued*

Compound Bond	$E(AB)$	R_o	δ_A	δ_B	t_i	E_c	E_i	E_{at} (kpm) Calc.	Exp.	$\Delta Hf°$ (kpm) Calc.	Exp.
SCl_2											
S'—Cl'	56.6	2.00	0.13	−0.06	0.09	52.3	14.9	134.4	129.5	−9.6	−4.7
SCl_4											
S'—Cl'	56.6	(2.00)	0.15	−0.04	0.09	52.3	14.9	268.8		−85.8	
S_2Br_2											
S'—S'	55.0	2.08	0.05	0.05	0	52.3	0	52.3			
S'—Br'	50.4	(2.16)	0.05	−0.05	0.05	48.3	7.7	112.0			
								164.3		+22.5	
SBr_2											
S'—Br'	50.4	(2.16)	0.06	−0.03	0.05	48.3	7.7	112.0		+8.2	
SI_2											
S'—I'	44.6	(2.37)	0.04	−0.02	0.03	43.3	4.2	95.0		+22.6	
SOF_2											
S'—F'	45.6	1.53	0.25	−0.12	0.18	42.8	39.1	163.8			
S'—O'''	73.9	1.41	0.25	−0.01	0.13	80.2	30.6	110.8			
								274.6	277.0	−110.6	−113.0
SO_2F_2											
S'—F'	45.6	1.53	0.25	−0.12	0.18	42.8	39.1	163.8			
S'—O'''	73.9	1.41	0.25	−0.01	0.13	80.2	30.6	221.6			
								385.4		−161.8	
$SOCl_2$											
S'—Cl'	56.6	2.07	0.16	−0.03	0.10	50.0	16.0	132.0			
S'—O'''	73.9	1.45	0.16	−0.09	0.12	78.9	27.5	106.4			
								238.4	235.2	−54.0	−50.8
SO_2Cl_2											
S'—Cl'	56.6	1.99	0.18	−0.02	0.10	52.0	17.7	139.4			
S'—O''	60.6	1.43	0.18	−0.07	0.13	64.9	30.2	190.2			
								329.6	331.0	−85.6	−87.0
$SOBr_2$											
S'—O''	60.6	(1.45)	0.11	−0.13	0.12	64.8	27.5	92.3			
S'—Br'	50.5	(2.18)	0.11	0.01	0.05	48.0	7.6	111.2			
								203.5	201.4	−23.9	−21.8
SeO_2											
Se'—O'''	66.2	1.61	0.15	−0.08	0.11	68.9	22.7	91.6			
Se''=O''	57.3	1.61	0.15	−0.08	0.11	89.3	34.0	123.3			
								214.9		−46.5	
$(SeO_2)_x$											
Se—O'	38.7	1.78	0.15	−0.08	0.11	36.4	20.5	113.8			
Se—O'''	66.2	1.73	0.15	−0.08	0.11	64.1	21.1	85.2			
								199.0		−30.6	

(*continued*)

ABLE A—*continued*

Compound Bond	$E(AB)$	R_o	δ_A	δ_B	t_i	E_c	E_i	E_{at} (kpm) Calc.	E_{at} (kpm) Exp.	$\Delta Hf°$ (kpm) Calc.	$\Delta Hf°$ (kpm) Exp.
$SeO_3)_x$											
Se—O′	38.7	(1.70)	0.17	− 0.06	0.11	36.4	20.5	113.8			
Se—O″	54.3	(1.70)	0.17	− 0.06	0.11	53.5	21.5	150.0			
								263.8		− 35.8	
eF_6											
Se′—F′	40.7	1.70	0.29	− 0.05	0.17	37.2	33.2	422.4	430.	− 259.8	− 267.
eCl_2											
Se—Cl′	50.5	(2.15)	0.10	− 0.05	0.08	46.5	12.4	117.8	115.0	− 10.4	− 7.6
eBr_2											
Se—Br′	45.1	(2.30)	0.05	− 0.02	0.03	43.8	4.3	96.2	107.6	+ 6.4	− 5.
eSe											
Te—Se	38.7	(2.50)	0.08	− 0.08	0.08	35.7	10.6	46.3	53.1	+ 48.4	+ 53.1
eF_6											
Te—F′	35.8	1.84	0.44	− 0.07	0.26	29.7	46.9	459.6	474.	− 300.7	− 315.
lO											
Cl—O′	44.5	1.55	0.03	− 0.03	0.03	47.6	6.4	54.0	64.5	+ 34.7	+ 24.2
l_2O											
Cl—O′	44.5	1.70	0.02	− 0.04	0.03	43.4	5.9	98.6	98.6	+ 19.2	+ 19.2
ClO_4											
H—O	48.8	(0.95)	0.33	− 0.08	0.20	42.8	69.9	112.7			
Cl—OH	44.5	1.64	− 0.02	− 0.08	0.03	45.0	6.1	51.1			
Cl—O′	44.5	1.42	− 0.02	− 0.08	0.03	52.0	7.0	177.0			
								340.8		− 21.2	(− 9.7)
lF											
Cl′—F′	46.9	1.63	0.09	− 0.09	0.09	44.5	18.3	62.8	59.9	− 14.8	− 11.9
lF_3											
Cl′—F′	46.9	1.70	0.13	− 0.04	0.08	43.1	15.6	58.7	(2 half-bonds)		
Cl′—F′	46.9	1.60	0.13	− 0.04	0.08	45.8	16.6	62.4			
								121.1	124.7	− 35.3	− 38.9
lF_5											
Cl—F′	46.9	1.62	0.15	− 0.03	0.09	44.8	18.4	63.2			
Cl—F′	46.9	1.72	0.15	− 0.03	0.09	42.2	17.4	119.2	(4 half-bonds)		
								182.4	181.4	− 58.8	− 57.8
lO_3F											
Cl—O′	44.5	(1.50)	0.07	0.01	0.03	49.2	6.6	167.4			
Cl—F′	47.0	(1.63)	0.07	− 0.10	0.09	44.6	18.3	62.9			
								230.3	231.3	− 4.7	− 5.7
rF											
Br′—F′	41.8	1.76	0.13	− 0.13	0.13	38.2	24.5	62.7	59.6	− 17.1	− 14.0
rF_3											
Br′—F′	41.8	1.81	0.20	− 0.07	0.13	37.2	23.8	61.0	(2 half-bonds)		

(*continued*)

TABLE A—*continued*

Compound Bond	E(AB)	R_o	δ_A	δ_B	t_i	E_c	E_i	E_{at} (kpm)		$\Delta Hf°$ (kpm)	
								Calc.	Exp.	Calc.	Exp.
Br″—F″	56.5	1.72	0.20	− 0.07	0.13	56.2	25.1	81.3			
								142.3	144.5	− 58.9	− 61.1
BrF₅											
Br‴—F‴	68.1	1.68	0.22	− 0.04	0.13	73.2	25.7	98.9			
Br′—F′	41.8	1.78	0.22	− 0.04	0.13	37.8	24.2	124.0	(4 half-bonds)		
								222.9	223.7	− 101.7	− 102.5
BrCl											
Br′—Cl′	51.7	2.14	0.04	− 0.04	0.04	49.4	6.2	55.6	51.3	+ 0.2	+ 3.5
IF											
I′—F′	37.0	1.91	0.21	− 0.21	0.21	31.2	36.5	67.7	67.0	− 23.3	− 22.6
IF₅											
I‴—F‴	66.4	1.75	0.37	− 0.07	0.22	60.4	41.7	102.1			
I′—F′	37.0	1.86	0.37	− 0.07	0.22	31.7	39.2	212.7	(2 full, 2 half-bonds)		
								314.8	320.0	− 194.8	− 200.0
IF₇											
I‴—F‴	66.4	1.75	0.39	− 0.06	0.22	60.4	41.7	102.1			
I′—F′	37.0	1.86	0.39	− 0.06	0.22	31.6	39.3	283.8	(2 full, 4 half-bonds)		
								385.9	386.9	− 228.1	− 229.1
ICl											
I′—Cl′	45.9	2.30	0.12	− 0.12	0.12	40.7	17.3	58.0	50.3	− 3.4	+ 4.3
IBr											
I′—Br′	40.8	(2.45)	0.08	− 0.08	0.08	37.8	10.8	48.6	42.4	+ 3.6	+ 9.8
TiF₄											
Ti—F″	30.9	1.80	1.12	− 0.28	0.70	10.5	129.1	558.4	558.8	− 370.3	− 370.7
TiCl₄											
Ti—Cl′	28.4	2.19	0.96	− 0.24	0.60	12.0	91.0	412.0	411.3	− 183.1	− 182.4
TiBr₄											
Ti—Br′	25.2	2.31	0.84	− 0.21	0.53	12.6	76.2	355.2	350.0	− 135.9	− 130.7
TiI₄											
Ti—I′	22.4	2.55	0.68	− 0.17	0.43	13.3	56.0	277.2	283.0	− 62.7	− 68.5
XeO₃											
Xe—O′	35.9	1.76	0.20	− 0.07	0.13	36.0	24.5	90.9	(3 half-bonds)		
									83.	+ 87.9	+ 96.
XeF₂											
Xe—F′	37.9	(1.98)	0.25	− 0.13	0.19	31.4	31.8	63.2	(2 half-bonds)		
									63.7	− 25.4	− 25.9
XeF₄											
Xe—F′	37.9	1.95	0.31	− 0.08	0.20	31.4	34.0	130.8	(4 half-bonds)		
									128.	− 55.2	− 52.

Miscellaneous Notes on Table B

All substances in Table B are in the gaseous state.

CHF_3, $CHCl_3$, HCHO, HCOOH

The treatment of C—H bonds between positively charged atoms of carbon and hydrogen is uncertain. There is the possibility of additional energy associated with internal protonic bridging. The latter may be responsible for the excess of experimental over calculated energy in CHF_3. Other evidences suggest that three chlorine atoms attached to the same carbon atom crowd one another as in CCl_4, although two chlorine atoms can be accommodated. This may account for the experimental value for $CHCl_3$ being lower than the calculated value. Errors in C—H bond energy calculation may be involved in all four of these compounds.

CCl_3CHO

Here is additional evidence of bond weakening through crowding of three chlorine atoms on one carbon.

$CH_2{=}CCl_2$, $CF_2{=}CCl_2$, $CCl_2{=}CCl_2$

For each of these, the calculated atomization energy is too high, for $CCl_2{=}CCl_2$ by approximately 34 kpm, and by 17 kpm for $CF_2{=}CCl_2$. These results suggest the possibility that for reasons as yet not understood, two chlorine atoms cannot be accommodated as readily on a carbon atom that forms a double bond as on a carbon that forms two other single bonds.

TABLE B

ATOMIZATION ENERGY DATA AND HEATS OF FORMATION OF SOME ORGANIC COMPOUNDS

Compound Bond	$E(AB)$	R_o	δ_A	δ_B	t_i	E_c	E_i	E_{at} (kpm)		ΔHf° (kpm)	
								Calc.	Exp.	Calc.	Exp.
CHF$_3$											
H—C	76.9	1.09	0.32	0.25	0.03	74.6	9.1	83.7			
C—F'	56.1	1.33	0.25	− 0.19	0.22	48.7	54.9	103.6			
C—F"	75.8	1.33	0.25	− 0.19	0.22	65.8	54.9	241.4			
								428.7	444.6	− 148.6	− 164.5
CHCl$_3$											
H—C	82.8	1.07	0.21	0.15	0.03	81.8	9.3	91.1			
C—Cl'	69.6	1.76	0.15	− 0.12	0.13	60.6	24.5	255.3			
								346.4	334.	− 35.7	− 24.
HCHO											
H—C	88.0	1.06	0.11	0.04	0.03	87.8	9.4	194.4			
C=O"	74.5	1.21	0.04	− 0.26	0.15	117.1	61.7	178.8			
								373.2	363.1	− 38.1	− 28.
HCOOH											
H—C	85.4	1.08	0.16	0.10	0.03	83.6	9.2	92.8			
C—O"	74.5	1.31	0.10	− 0.21	0.16	71.2	40.5	111.7			
C=O"	74.5	1.25	0.10	− 0.21	0.16	112.0	63.8	175.8			
O—H	54.6	0.95	− 0.21	0.16	0.19	48.4	66.4	114.8			
								495.1	481.4	− 79.4	− 86.7
CH$_2$F$_2$											
H—C	82.8	1.09	0.21	0.14	0.03	80.3	9.1	178.8			
C—F"	75.8	1.36	0.14	− 0.28	0.21	65.2	51.3	233.0			
								411.8	420.1	− 98.5	− 106.3
CH$_2$Cl$_2$											
H—C	86.3	1.09	0.14	0.08	0.03	83.7	9.1	185.6			
C—Cl'	69.6	1.77	0.08	− 0.18	0.13	60.2	24.4	169.2			
								354.8	355.	− 21.1	− 21
HCONH$_2$											
H—C	88.0	(1.08)	0.11	0.05	0.03	86.2	9.2	95.4			
H—N	60.3	(1.02)	0.11	− 0.12	0.11	55.8	35.8	183.2			
C—N"	74.6	1.30	0.05	− 0.12	0.08	79.7	20.4	100.1			
C=O"	74.5	1.25	0.05	− 0.26	0.15	113.2	59.8	173.0			
								551.7	546.7	− 51.5	− 46.5
CH$_3$Cl											
H—C	89.8	1.09	0.07	0.01	0.03	(3 × 98.7)		296.1			
C—Cl'	69.6	1.78	0.01	− 0.24	0.12	60.6	22.4	83.0			
								379.1	376.3	− 22.4	− 19.6

(continued)

ABLE B—*continued*

ompound Bond	$E(AB)$	R_o	δ_A	δ_B	t_i	E_c	E_i	E_{at} (kpm) Calc.	E_{at} (kpm) Exp.	$\Delta Hf°$ (kpm) Calc.	$\Delta Hf°$ (kpm) Exp.
H₃Br											
C—H	(98.7)							296.1			
C—Br′	62.0	1.94	0.00	−0.17	0.08	56.2	13.7	69.9			
								366.0	362.7	−11.7	−8.4
CH₃I											
H—C	91.8	1.09	0.03	−0.03	0.03	(3 × 98.7)		296.1			
C—I′	54.8	2.14	−0.03	−0.05	0.01	53.3	1.6	54.9			
								351.0	348.2	+2.1	+4.9
CH₄											
H—C	92.8	1.09	0.01	−0.05	0.03	90.0	9.1	396.4	397.6	−16.7	−17.9
CH₃OH											
H—C	89.8	1.09	0.07	0.01	0.03	(3 × 98.7)		296.1			
C—O′	52.4	1.43	0.01	−0.29	0.15	46.4	34.8	81.2			
O—H	57.5	0.96	−0.29	0.07	0.18	51.1	62.3	113.4			
								490.7	487.4	−51.4	−48.1
CH₃SH											
H—C	91.8	1.09	0.03	−0.03	0.03	(3 × 98.7 =)		296.1			
C—S′	67.1	1.82	−0.03	−0.10	0.04	64.0	7.3	71.3			
H—S	74.6	1.34	0.03	−0.10	0.07	70.4	17.3	87.7			
								455.1	449.3	−8.8	−3.0
CH₃NH₂											
H—C	91.3	1.09	0.04	−0.02	0.03	(3 × 98.7 =)		296.1			
H—N	62.7	1.01	0.04	−0.18	0.11	58.6	36.2	189.6			
C—N′	57.2	1.47	−0.02	−0.18	0.08	54.1	18.1	72.2			
								557.9	551.5	−13.1	−6.7
CCl₃CHO											
H—C	82.2	1.09	0.22	0.15	0.03	79.7	9.1	88.8			
C—C	83.2	1.52	0.15	0.15	0	71.7	0	71.7			
C—Cl′	69.6	1.76	0.15	−0.12	0.13	60.5	24.6	255.3			
C=O″	74.5	1.15	0.15	−0.17	0.16	121.7	69.3	191.0			
								606.8	588.6	−65.2	−47.0
C₂H₂											
H—C	91.8	1.06	0.03	−0.03	0.03	91.6	9.4	202.0			
C≡C	83.2	1.20	−0.03	−0.03	0	189.0	0	189.0			
								391.0	392.6	+55.8	+54.2
CH₂CO											
H—C	88.4	1.08	0.10	0.04	0.03	(98.7 × 2 =)		197.4			
C=C	83.2	1.31	0.04	0.04	0	140.8	0	140.8			
C=O″	74.5	1.17	0.04	−0.27	0.15	121.0	63.8	184.8			
								523.0	521.0	−16.6	−14.6

(continued)

TABLE B—*continued*

Compound Bond	$E(AB)$	R_o	δ_A	δ_B	t_i	E_c	E_i	E_{at} (kpm) Calc.	E_{at} (kpm) Exp.	$\Delta Hf°$ (kpm) Calc.	$\Delta Hf°$ (kpm) Exp
$C_2H_2F_2$											
H—C		(1.08)	0.18	0.12	0.03	$(2 \times 98.7 =)$		197.4			
C=C	83.2	(1.31)	0.12	0.12	0	146.7	0	146.7			
C—F′	56.1	(1.31)	0.12	−0.30	0.21	50.1	53.2	206.6			
								550.7	563.2	−66.1	−78.(
$C_2H_2Cl_2$											
H—C						$(2 \times 98.7 =)$		197.4			
C=C	83.2	(1.36)	0.06	0.06	0	141.2	0	141.2			
C—Cl′	69.6	1.69	0.06	−0.19	0.13	63.1	25.5	177.2			
								515.8	504.4	−10.8	+0.(
CH_3CN											
H—C	90.3	1.10	0.06	0.00	0.03	$(3 \times 98.7 =)$		296.1			
C—C	83.2	1.46	0	0	0	87.8		87.8			
C=N‴	89.0	1.10	0.00	−0.16	0.08	168.6	36.2	204.8			
								588.7	590.9	+23.2	+21.0
CH_3NC											
H—C	90.3	1.10	0.06	0.00	0.03	$(3 \times 98.7 =)$		296.1			
C—N″	74.6	(1.45)	0.0	−0.16	0.08	71.5	18.3	89.8			
C=N‴	89.0	1.16	0.0	−0.16	0.08	160.0	34.3	194.3			
								580.2	576.0	+31.7	+35.9
CH_2CHCl											
H—C	89.8	1.07	0.07	0.01	0.03	88.7	9.3	98.0			
H—C						$(2 \times 98.7 =)$		197.4			
C=C	83.2	1.34	0.01	0.01	0	142.0	0	142.0			
C—Cl′	69.6	1.69	0.01	−0.24	0.12	63.4	23.6	87.0			
								524.4	520.5	+3.6	+7.5
CH_3COCl											
H—C						$(3 \times 98.7 =)$		296.1			
C—C	83.2	1.50	0.12	0.12	0	75.2	0	75.2			
C—Cl′	69.6	(1.72)	0.12	−0.14	0.13	62.3	25.1	87.4			
C=O″	74.5	1.17	0.12	−0.20	0.16	119.5	68.2	187.7			
								646.4	645.8	−58.8	−58.2
C_2H_4											
H—C	92.2	(1.09)	0.02	−0.04	0.03	$(4 \times 98.7 =)$		394.8			
C=C	83.2	1.33	−0.04	−0.04	0	144.5	0	144.5			
								539.3	538.5	+11.7	+12.5
CH_3CHO											
H—C	89.8	1.09	0.07	0.01	0.03	87.1	9.1	96.2			
H—C						$(3 \times 98.7 =)$		296.1			
C—C	83.2	1.50	0.01	0.01	0	84.6	0	84.6			
C=O″	74.5	1.22	0.01	−0.29	0.15	116.0	61.2	177.2			
								654.1	650.4	−43.5	−39.8

(*continued*)

TABLE B—*continued*

Compound Bond	$E(AB)$	R_o	δ_A	δ_B	t_i	E_c	E_i	E_{at} (kpm) Calc.	E_{at} (kpm) Exp.	$\Delta Hf°$ (kpm) Calc.	$\Delta Hf°$ (kpm) Exp.
CH_3COOH											
H—C	88.0	1.08	0.11	0.05	0.03	$(3 \times 98.7 =)$		296.1			
C—C	83.2	1.50	0.05	0.05	0	81.1	0	81.1			
C—O″	74.5	1.31	0.05	− 0.26	0.15	72.0	38.0	110.0			
C=O″	74.5	1.25	0.05	− 0.26	0.15	113.2	59.8	173.0			
O—H	56.3	0.95	0.11	− 0.26	0.18	50.5	62.9	113.4			
								773.6	774.9	− 103.4	− 104.7
$HCOOCH_3$											
H—C	88.0	1.09	0.11	0.05	0.03	85.4	9.1	94.5			
H—C						$(3 \times 98.7 =)$		296.1			
C—O′	52.4	1.44	0.05	− 0.26	0.15	46.1	34.6	80.7			
C—O″	74.5	1.33	0.05	− 0.26	0.15	70.9	37.4	108.3			
C=O″	74.5	1.20	0.05	− 0.26	0.15	117.9	62.3	180.2			
								759.8	753.8	− 89.6	− 83.6
CH_2ICH_2I											
H—C						$(4 \times 98.7 =)$		394.8			
C—C	83.2	1.54	− 0.01	− 0.01	0	83.2	0	83.2			
C—I′	54.8	(2.18)	− 0.01	− 0.04	0.01	52.3	1.5	107.6			
								585.6	586.1	+ 16.4	+ 15.9
C_2H_5ONO											
H—C	88.0	(1.09)	0.11	0.05	0.03	$(5 \times 98.7 =)$		493.5			
C—C	83.2	(1.54)	0.05	0.05	0	79.0	0	79.0			
C—O′	52.4	1.47	0.05	− 0.26	0.15	45.1	33.9	79.0			
N′—O″	51.2	(1.44)	− 0.12	− 0.26	0.07	48.3	16.1	64.4			
N″=O″	66.8	(1.19)	− 0.12	− 0.26	0.07	114.3	29.3	143.6			
								859.5	860.1	− 24.2	− 24.8
C_2H_5Cl											
H—C	90.8	1.10	0.05	− 0.01	0.03	$(5 \times 98.7 =)$		493.5			
C—C	83.2	(1.55)	− 0.01	− 0.01	0	82.7	0	82.7			
C—Cl′	69.6	1.78	− 0.01	− 0.25	0.12	60.6	22.4	83.0			
								659.2	657.3	− 27.0	− 25.1
C_2H_5Br											
H—C	91.3	1.11	0.04	− 0.02	0.03	$(5 \times 98.7 =)$		493.5			
C—C	83.2	1.55	− 0.02	− 0.02	0	82.7	0	82.7			
C—Br	62.0	1.94	− 0.02	− 0.17	0.08	56.2	13.7	69.9			
								646.1	642.8	− 16.3	− 13.0
C_2H_5I											
H—C						$(5 \times 98.7 =)$		493.5			
C—C	83.2	1.54	− 0.04	− 0.04	0	83.2	0	83.2			
C—I′	54.8	2.18	− 0.04	− 0.05	0.01	52.3	1.5	53.8			
								630.5	630.4	− 1.9	− 1.8

(*continued*)

TABLE B—*continued*

Compound Bond	$E(AB)$	R_o	δ_A	δ_B	t	E_c	E_i	E_{at} (kpm) Calc.	E_{at} (kpm) Exp.	$\Delta Hf°$ (kpm) Calc.	$\Delta Hf°$ (kpm) Exp
C_2H_6											
H—C	92.8	1.09	0.02	− 0.04	0.03	$(6 \times 98.7 =)$		592.2			
C—C	83.2	1.54	− 0.04	− 0.04	0	83.2	0	83.2			
								675.4	675.4	− 20.2	− 20.
C_2H_5OH											
H—C	90.8	1.09	0.05	− 0.01	0.03	$(5 \times 98.7 =)$		493.5			
C—C	83.2	1.55	− 0.01	− 0.01	0	82.7	0	82.7			
C—O′	52.4	1.43	− 0.01	− 0.31	0.15	46.4	34.8	81.2			
O—H	58.1	0.96	− 0.31	0.05	0.18	51.6	62.3	113.9			
								771.3	770.8	− 56.5	− 56.
CH_3OCH_3											
H—C	90.8	1.09	0.05	− 0.01	0.03	$(6 \times 98.7 =)$		592.2			
C—O′	52.4	1.42	− 0.01	− 0.31	0.15	46.7	35.1	163.6			
								755.8	759.1	− 41.0	− 44.
$(CH_2OH)_2$											
H—C	88.9	1.08	0.08	0.02	0.03	$(4 \times 98.7 =)$		394.8			
C—C	83.2	1.53	0.02	0.02	0	82.1	0	82.1			
C—O′	52.4	1.43	0.02	− 0.28	0.15	46.4	34.8	162.4			
O—H	56.8	0.97	− 0.28	0.08	0.18	50.0	61.6	223.2			
								862.5	866.9	− 88.1	− 92.
$(CH_3)_2S$											
H—C	91.8	(1.09)	0.03	− 0.03	0.03	$(6 \times 98.7 =)$		592.2			
C—S	67.1	1.82	− 0.03	− 0.11	0.04	64.1	7.3	142.8			
								735.0	728.7	− 13.2	− 6.
C_2H_5SH											
						$(5 \times 98.7 =)$		493.5			
C—C	83.2	1.54	− 0.03	− 0.03	0	83.2	0	83.2			
C—S	67.1	(1.81)	− 0.03	− 0.11	0.04	64.5	7.3	71.8			
S—H	74.6	(1.33)	− 0.11	0.03	0.07	70.9	17.5	88.4			
								736.9	732.8	− 15.1	− 11.
$(CH_3)_2SO$											
H—C						$(6 \times 98.7 =)$		592.2			
C—S	67.7	1.84	0.0	− 0.08	0.04	63.9	7.2	142.2			
S—O″	60.6	1.47	− 0.08	− 0.30	0.11	64.6	24.8	89.4			
								823.8	817.4	− 42.4	− 36.
$(CH_3)_2SO_2$											
H—C						$(6 \times 98.7 =)$		592.2			
C—S	67.7	(1.80)	0.03	− 0.05	0.04	65.4	7.4	145.6			
S′—O″	60.6	1.43	− 0.05	− 0.28	0.11	66.4	25.5	183.8			
								921.6	929.7	− 80.6	− 88.

(continued)

TABLE B—*continued*

Compound Bond	E(AB)	R_o	δ_A	δ_B	t_i	E_c	E_i	E_{at} (kpm) Calc.	E_{at} (kpm) Exp.	$\Delta Hf°$ (kpm) Calc.	$\Delta Hf°$ (kpm) Exp.
(CH₃)₂NH											
H—C	91.8	1.08	0.03	− 0.03	0.03	(6 × 98.7 =)		592.2			
C—N′	57.2	1.46	− 0.03	− 0.18	0.08	54.4	18.2	145.2			
N—H	58.7	(1.02)	− 0.18	0.03	0.11	54.3	35.8	90.1			
								827.5	826.9	− 7.2	− 6.6
C₂H₅NH₂											
H—C	91.8	1.09	0.03	− 0.03	0.03	(5 × 98.7 =)		493.5			
C—C	83.2	1.54	− 0.03	− 0.03	0	83.2	0	83.2			
C—N′	57.2	1.47	− 0.03	− 0.18	0.08	54.1	18.1	72.2			
N—H	58.7	(1.02)	− 0.18	0.03	0.11	54.3	35.8	180.2			
								829.1	831.9	− 8.8	− 11.6
CF₂CCl₂											
C=C	83.2	(1.32)	0.24	0.24	0	145.6	0	145.6			
C—Cl′	69.6	(1.72)	0.24	− 0.04	0.14	61.3	27.0	176.6			
C—F′	56.1	(1.31)	0.24	− 0.20	0.22	49.4	55.7	210.2			
								532.4	514.9	− 93.8	− 76.3
CCl₂CCl₂											
C=C	83.2	1.30	0.18	0.18	0	147.8	0	147.8			
C—Cl′	69.6	1.72	0.18	− 0.09	0.13	62.0	25.1	348.4			
								496.2	461.9	− 37.2	− 2.9
CH₃CCH											
H—C	91.8	1.06	0.03	− 0.03	0.03	91.6	9.4	101.0			
H—C						(3 × 98.7 =)		296.1			
C—C	83.2	1.46	− 0.03	− 0.03	0	87.8	0	87.8			
C≡C	83.2	1.21	− 0.03	− 0.03	0	187.4	0	187.4			
								672.3	678.0	+ 50.0	+ 44.3
CH₂=CHCHO											
H—C	89.8	1.09	0.07	0.01	0.03	87.1	9.1	192.4			
H—C						(2 × 98.7 =)		197.4			
C—C	83.2	1.45	0.01	0.01	0	87.5	0	87.5			
C=C	83.2	1.36	0.01	0.01	0	139.9	0	139.9			
C=O″	74.5	1.22	0.01	− 0.29	0.15	116.0	61.2	177.2			
								794.4	802.4	− 12.5	− 20.5
C₃H₆											
H—C	92.2	1.08	0.02	− 0.04	0.03	90.3	9.2	99.5			
H—C						(5 × 98.7 =)		493.5			
C—C	83.2	1.49	− 0.04	− 0.04	0	86.0	0	86.0			
C=C	83.2	1.35	− 0.04	− 0.04	0	142.4	0	142.4			
								821.4	821.6	+ 5.1	+ 4.9
CH₃COCH₃											
H—C	90.8	1.09	0.05	− 0.01	0.03	(6 × 98.7 =)		592.2			

(continued)

206

TABLE B—*continued*

Compound Bond	$E(AB)$	R_o	δ_A	δ_B	t_i	E_c	E	E_{at} (kpm) Calc.	E_{at} (kpm) Exp.	$\Delta Hf°$ (kpm) Calc.	$\Delta Hf°$ (kpm) Exp.
C—C	83.2	1.52	− 0.01	− 0.01	0	84.3	0	168.6			
C=O″	74.5	1.24	− 0.01	− 0.31	0.15	114.1	58.9	173.0			
								933.8	938.5	− 47.7	− 52.4
C_2H_5CHO											
H—C	90.8	1.09	0.05	− 0.01	0.03	88.1	9.1	97.2			
H—C						$(5 \times 98.7 =)$		493.5			
C—C	83.2	1.52	− 0.01	− 0.01	0	84.3	0	168.6			
C=O″	74.5	1.22	− 0.01	− 0.31	0.15	116.0	61.2	177.2			
								936.5	935.3	− 50.4	− 49.2
C_2H_5COOH											
H—C	89.4	(1.09)	0.08	0.02	0.03	$(5 \times 98.7 =)$		493.5			
C—C	83.2	(1.52)	0.02	0.02	0	82.6	0	165.2			
C—O″	74.5	(1.31)	0.02	− 0.28	0.15	72.0	38.0	110.0			
C=O″	74.5	(1.25)	0.02	− 0.28	0.15	113.3	59.8	173.1			
O—H	57.1	(0.95)	− 0.28	0.08	0.18	51.3	62.9	114.2			
								1056.0	1054.5	− 110.3	− 108.8
$CH_2=CHCH_2NH_2$											
H—C	91.3	(1.08)	0.04	− 0.02	0.03	89.4	9.2	98.6			
H—C						$(4 \times 98.7 =)$		394.8			
C—C	83.2	(1.52)	− 0.02	− 0.02	0	84.3	0	84.3			
C=C	83.2	(1.33)	− 0.02	− 0.02	0	144.5	0	144.5			
C—N′	57.2	(1.47)	− 0.02	− 0.18	0.08	54.1	18.1	72.2			
N—H	62.6	(1.02)	− 0.18	0.04	0.11	57.9	35.8	187.4			
								981.8	984.8	+ 9.8	+ 6.8
C_3H_7Cl											
H—C	91.3	1.09	0.04	− 0.02	0.03	$(7 \times 98.7 =)$		690.9			
C—C	83.2	1.54	− 0.02	− 0.02	0	83.2	0	166.4			
C—Cl	69.6	1.75	− 0.02	− 0.26	0.12	61.6	22.8	84.4			
								941.7	938.6	− 34.0	− 30.9
C_3H_7Br											
H—C	91.3	(1.09)	0.04	− 0.02	0.03	$(7 \times 98.7 =)$		690.9			
C—C	83.2	(1.54)	− 0.02	− 0.02	0	83.2	0	166.4			
C—Br	62.0	(1.91)	− 0.02	− 0.19	0.08	57.0	13.9	70.9			
								928.2	929.3	− 22.9	− 24.0
C_3H_7OH											
H—C	90.8	1.09	0.04	− 0.02	0.03	$(7 \times 98.7 =)$		690.9			
C—C	83.2	1.54	− 0.02	− 0.02	0	83.2	0	166.4			
C—O′	52.4	1.45	− 0.02	− 0.31	0.15	45.8	34.3	80.1			
O—H	58.1	0.96	− 0.31	0.04	0.18	51.6	62.3	113.9			
								1051.3	1051.5	− 61.0	− 61.2
$CH_3OC_2H_5$											
H—C	90.8	(1.09)	0.04	− 0.02	0.03	$(8 \times 98.7 =)$		789.6			

(continued)

TABLE B—*continued*

Compound Bond	$E(AB)$	R_o	δ_A	δ_B	t_i	E_c	E_i	E_{at} (kpm) Calc.	E_{at} (kpm) Exp.	$\Delta Hf°$ (kpm) Calc.	$\Delta Hf°$ (kpm) Exp.
C—C	83.2	(1.54)	−0.02	−0.02	0	83.2	0	83.2			
C—O′	52.4	(1.45)	−0.02	−0.31	0.15	45.8	34.3	160.2			
								1033.0	1042.0	−42.7	−51.7
(CH₃)₃N											
H—C	91.8	1.09	0.03	−0.03	0.03	(9 × 98.7 =)		888.3			
C—N′	57.2	1.47	−0.03	−0.19	0.08	54.1	18.1	216.6			
								1104.9	1099.8	−9.1	−4.0
CHCC₂H₅											
H—C	92.2	1.09	0.02	−0.04	0.03	92.0	9.4	101.4			
H—C						(5 × 98.7 =)		493.5			
C—C	83.2	1.47	−0.04	−0.04	0	87.2	0	174.4			
C≡C	83.2	1.20	−0.04	−0.04	0	189.0	0	189.0			
								958.3	959.8	+39.5	+38
C₄H₆											
H—C	92.2	1.08	0.02	−0.04	0.03	90.3	9.2	199.0			
H—C						(4 × 98.7 =)		394.8			
C=C	83.2	1.34	−0.04	−0.04	0	143.4	0	286.8			
C—C	83.2	1.48	−0.04	−0.04	0	86.6	0	86.6			
								967.2	970.3	+30.6	+27.5
CH₂CHCOOCH₃											
H—C	89.4	(1.08)	0.08	0.02	0.03	87.5	9.2	96.7			
H—C						(5 × 98.7 =)		493.5			
C—C	83.2	(1.44)	0.02	0.02	0	87.2	0	87.2			
C=C	83.2	(1.34)	0.02	0.02	0	140.6	0	140.6			
C—O′	52.4	(1.46)	0.02	−0.28	0.15	45.5	34.1	79.6			
C—O″	74.5	1.36	0.02	−0.28	0.15	70.3	36.6	106.9			
C=O″	74.5	1.22	0.02	−0.28	0.15	117.6	61.2	178.8			
								1183.3	1187.1	−66.3	−70.1
(CH₃CO)₂O											
H—C	88.4	(1.09)	0.10	0.04	0.03	(6 × 98.7 =)		592.2			
C—C	83.2	(1.50)	0.04	0.04	0	85.4	0	170.8			
C—O″	74.5	1.31	0.04	−0.26	0.15	72.0	38.0	220.0			
C=O″	74.5	1.25	0.04	−0.26	0.15	113.2	59.8	346.0			
								1329.0	1325.4	−152.4	−148.8
2-C₄H₈											
H—C	92.2	(1.09)	0.02	−0.04	0.03	89.4	9.1	197.0			
H—C						(6 × 98.7 =)		592.2			
C—C	83.2	1.52	−0.04	−0.04	0	84.3	0	168.6			
C=C	83.2	1.34	−0.04	−0.04	0	143.4	0	143.4			
								1101.2	1103.	+0.4	−1

(*continued*)

TABLE B—*continued*

Compound Bond	$E(AB)$	R_o	δ_A	δ_B	t_i	E_c	E_i	E_{at} (kpm) Calc.	Exp.	$\Delta Hf°$ (kpm) Calc.	Ex
C₃H₇CHO											
H—C	90.8	(1.09)	0.05	− 0.01	0.03	88.1	9.1	97.2			
H—C						(7 × 98.7 =)		690.9			
C—C	83.2	(1.54)	− 0.01	− 0.01	0	83.2	0	249.6			
C=O″	74.5	(1.22)	− 0.01	− 0.31	0.15	116.0	61.2	177.2			
								1214.9	1214.0	− 53.3	− 52
CH₃COOC₂H₅											
H—C	89.8	1.09	0.07	0.01	0.03	(8 × 98.7 =)		789.6			
C—C	83.2	1.52	0.01	0.01	0	83.4	0	166.8			
C—O′	52.4	1.46	0.01	− 0.29	0.15	45.5	34.1	79.6			
C—O″	74.5	1.31	0.01	− 0.29	0.15	72.0	38.0	110.0			
C=O″	74.5	1.25	0.01	− 0.29	0.15	113.2	59.8	173.0			
								1319.0	1323.2	− 97.8	− 102
C₄H₁₀											
H—C	92.8	1.09	0.01	− 0.05	0.03	(10 × 98.7 =)		987.0			
C—C	83.2	1.54	− 0.05	− 0.05	0	83.2	0	249.6			
								1236.6	1236.8	− 30.4	− 30
C₄H₉OH											
H—C	92.2	(1.09)	0.04	− 0.02	0.03	(9 × 98.7 =)		888.3			
C—C	83.2	(1.54)	− 0.02	− 0.02	0	83.2	0	249.6			
C—O′	52.4	(1.44)	− 0.02	− 0.32	0.15	46.1	34.6	80.7			
O—H	59.1	(0.96)	− 0.32	0.04	0.18	52.5	62.3	114.8			
								1333.4	1334.	− 67.6	− 68
(C₂H₅)₂O											
H—C	91.3	(1.09)	0.04	− 0.02	0.03	(10 × 98.7 =)		987.0			
C—C	83.2	1.50	− 0.02	− 0.02	0	85.4	0	170.8			
C—O′	52.4	1.43	− 0.02	− 0.32	0.15	46.4	34.8	162.4			
								1320.2	1326.1	− 54.4	− 60.
(C₂H₅)₂NH											
H—C						(10 × 98.7 =)		987.0			
C—C	83.2	(1.54)	− 0.03	− 0.03	0	83.2	0	166.4			
C—N′	57.2	(1.47)	− 0.03	− 0.19	0.08	54.1	18.1	144.4			
N—H	63.0	(1.01)	− 0.19	0.03	0.11	58.8	36.2	95.0			
								1392.8	1388.4	− 21.5	− 17.
Cyc-C₅H₆											
H—C	91.8	1.08	0.03	− 0.03	0.03	(6 × 98.7 =)		592.2			
C—C	83.2	1.53	− 0.03	− 0.03	0	83.7	0	167.4			
=C—C=	83.2	1.46	− 0.03	− 0.03	0	87.8	0	87.8			
C=C	83.2	1.35	− 0.03	− 0.03	0	142.4	0	284.8			
								1132.2	1136.5	+ 36.9	+ 32.
CHCC₃H₇											
H—C	92.2	1.07	0.02	− 0.04	0.03	91.1	9.3	100.4			

(*continued*

TABLE B—*continued*

Compound Bond	$E(AB)$	R_o	δ_A	δ_B	t_i	E_c	E_i	E_{at} (kpm) Calc.	Exp.	$\Delta Hf°$ (kpm) Calc.	Exp.
H—C						$(7 \times 98.7 =)$		690.9			
C—C	83.2	(1.54)	−0.04	−0.04	0	83.2	0	166.4			
C—C≡	83.2	1.46	−0.04	−0.04	0	87.8	0	87.8			
C≡C	83.2	(1.21)	−0.04	−0.04	0	187.4	0	187.4			
								1232.9	1238.8	+40.4	+34.5
Cyc-C_5H_{10}											
H—C	92.2	(1.09)	0.02	−0.04	0.03	$(10 \times 98.7 =)$		987.0			
C—C	83.2	1.54	−0.04	−0.04	0	83.2	0	416.0			
								1403.0	1395.8	−25.5	−18.3
1-C_5H_{10}											
H—C	92.2	(1.09)	0.02	−0.04	0.03	$(10 \times 98.7 =)$		987.0			
C—C	83.2	(1.54)	−0.04	−0.04	0	83.2	0	249.6			
C=C	83.2	(1.34)	−0.04	−0.04	0	143.4	0	143.4			
								1380.0	1382.5	−2.5	−5.0
$C_2H_5COOC_2H_5$											
H—C	90.3	(1.09)	0.06	0	0.03	$(10 \times 98.7 =)$		987.0			
C—C	83.2	(1.52)	0	0	0	84.3	0	252.9			
C—O′	52.4	(1.46)	0	−0.30	0.15	45.5	34.1	79.6			
C—O″	74.5	(1.31)	0	−0.30	0.15	72.0	38.0	110.0			
C=O″	74.5	(1.25)	0	−0.30	0.15	113.2	59.8	173.0			
								1602.5	1609.1	−105.8	−112.4
C_5H_{12}											
H—C	92.2	(1.09)	0.02	−0.04	0.03	$(12 \times 98.7 =)$		1184.4			
C—C	83.2	(1.54)	−0.04	−0.04	0	83.2	0	332.8			
								1517.2	1516.5	−35.5	−34.8
C_6H_6											
H—C	91.8	1.09	0.03	−0.03	0.03	$(6 \times 97.3 =)$		583.8			
C—C	83.2	1.39	−0.03	−0.03	0	122.6	0	735.6			
								1319.4	1320.6	+21.0	+19.8
C_6H_5OH											
H—C	90.3	(1.09)	0.06	0	0.03	$(5 \times 97.3 =)$		486.5			
C—C	83.2	(1.39)	0	0	0	122.6	0	735.6			
C—O′	52.4	(1.36)	0	−0.30	0.15	48.8	36.6	85.4			
O—H	57.7	(0.96)	−0.30	0.06	0.18	51.3	62.3	113.6			
								1421.1	1421.7	−21.1	−21.7
$C_6H_5NH_2$											
H—C	91.3	(1.09)	0.04	−0.02	0.03	$(5 \times 97.3 =)$		486.5			
C—C	83.2	(1.39)	−0.02	−0.02	0	122.6	0	735.6			
C—N′	57.2	(1.47)	−0.02	−0.18	0.08	54.1	18.1	72.2			
N—H	62.6	(1.02)	−0.18	0.04	0.11	58.0	35.8	187.6			
								1481.9	1485.1	+23.6	+20.4

(*continued*)

TABLE B—*continued*

Compound Bond	E(AB)	R_o	δ_A	δ_B	t_i	E_c	E_i	E_{at}(kpm) Calc.	E_{at}(kpm) Exp.	$\Delta Hf°$ (kpm) Calc.	$\Delta Hf°$ (kpm) Exp.
$(CH_2)_4(COOH)_2$											
H—C	88.9	(1.09)	0.09	0.03	0.03	$(8 \times 98.7 =)$		789.6			
C—C	83.2	1.51	0.03	0.03	0	84.9	0	424.5			
C—O″	74.5	1.29	0.03	− 0.27	0.15	73.1	38.6	223.4			
C=O″	74.5	1.23	0.03	− 0.27	0.15	115.0	60.7	351.4			
O—H	56.8	0.95	− 0.27	0.09	0.18	51.0	62.9	227.8			
								2016.7	2003.4	− 229.5	− 216.2
$Cyc\text{-}C_6H_{12}$											
H—C	92.2	1.09	0.02	− 0.04	0.03	$(12 \times 98.7 =)$		1184.4			
C—C	83.2	1.54	− 0.04	− 0.04	0	83.2	0	499.2			
								1683.6	1682.4	− 30.6	− 29.4
$1\text{-}C_6H_{12}$											
H—C	92.2	(1.09)	0.02	− 0.04	0.03	$(12 \times 98.7 =)$		1184.4			
C—C	83.2	(1.54)	− 0.04	− 0.04	0	83.2	0	332.8			
C=C	83.2	(1.34)	− 0.04	− 0.04	0	143.4	0	143.4			
								1660.6	1663.0	− 7.6	− 10.0
nC_6H_{14}											
H—C	92.2	(1.09)	0.02	− 0.04	0.03	$(14 \times 98.7 =)$		1380.8			
C—C	83.2	(1.54)	− 0.04	− 0.04	0	83.2	0	416.0			
								1796.8	1797.2	− 39.6	− 40.0
$(C_2H_5)_3N$											
H—C						$(15 \times 98.7 =)$		1480.5			
C—C	83.2	(1.54)	− 0.04	− 0.04	0	83.2	0	249.6			
C—N′	57.2	1.47	− 0.04	− 0.19	0.08	54.1	18.1	216.6			
								1946.7	1945.2	− 24.4	− 22.9
$(CH_2)_6(NH_2)_2$											
H—C	91.8	(1.09)	0.03	− 0.03	0.03	$(12 \times 98.7 =)$		1184.4			
C—C	83.2	1.53	− 0.03	− 0.03	0	83.7	0	418.5			
C—N′	57.2	1.51	− 0.03	− 0.18	0.08	52.6	17.6	140.4			
N—H	63.0	(1.02)	− 0.18	0.03	0.11	58.3	35.8	376.4			
								2119.7	2118.0	− 32.3	− 30.6
C_6H_5CHO											
H—C	90.3	(1.09)	0.06	0	0.03	87.6	9.1	96.7			
H—C						$(5 \times 97.3 =)$		486.5			
=C—C	83.2	(1.50)	0	0	0	85.4	0	85.4			
C—C	83.2	(1.39)	0	0	0	122.6	0	735.6			
C=O″	74.5	(1.22)	0	− 0.30	0.15	116.0	61.2	177.2			
								1581.4	1580.9	− 10.1	− 9.6
$CH_3C_6H_5$											
H—C	91.8	1.09	0.03	− 0.03	0.03	$(3 \times 98.7 =)$		296.1			
H—C						$(5 \times 97.3 =)$		486.5			

(continued)

TABLE B—*continued*

Compound Bond	E(AB)	R_o	δ_A	δ_B	t_i	E_c	E_i	E_{at} (kmp) Calc.	Exp.	$\Delta Hf°$ (kmp) Calc.	Exp.
C—CH₃	83.2	1.51	−0.03	−0.03	0	84.9	0	84.9			
C—C	83.2	1.39	−0.03	−0.03	0	122.6	0	735.6			
								1603.1	1603.7	+12.8	+12.2
1-C₇H₁₄											
H—C	92.2	(1.09)	0.02	−0.04	0.03	(14 × 98.7 =)		1381.8			
C—C	83.2	(1.54)	−0.04	−0.04	0	83.2	0	416.0			
C=C	83.2	(1.33)	−0.04	−0.04	0	144.5	0	144.5			
								1942.3	1943.4	−13.8	−14.9
*n*C₇H₁₆											
H—C	92.2	(1.09)	0.02	−0.04	0.03	(16 × 98.7 =)		1579.2			
C—C	83.2	(1.54)	−0.04	−0.04	0	83.2	0	499.2			
								2078.4	2077.6	−45.7	−44.9
C₆H₅CH=CH₂											
H—C	92.2	(1.09)	0.02	−0.04	0.03	(3 × 98.7 =)		296.1			
H—C						(5 × 97.3 =)		486.5			
C—C	83.2	1.50	−0.04	−0.04	0	85.4	0	85.4			
C=C	83.2	1.33	−0.04	−0.04	0	144.5	0	144.5			
C—C	83.2	1.39	−0.04	−0.04	0	122.6	0	735.6			
								1748.1	1752.6	+39.1	+34.6
Cyc-C₈H₈											
H—C	91.8	1.09	0.03	−0.03	0.03	(8 × 98.7 =)		789.6			
C—C	83.2	1.46	−0.03	−0.03	0	87.8	0	351.2			
C=C	83.2	1.33	−0.03	−0.03	0	144.5	0	578.0			
								1718.8	1717.1	+68.3	+70.

AUTHOR INDEX

SUBJECT INDEX

A

Accuracy, of bond energy calculation, 24
Acetaldehyde, 202
Acetic acid, 203
  atomization energy of, 149, 203
  protonic bridging in, 115
Acetic anhydride, 151, 207
Acetone, 151, 205
Acetonitrile, 151, 202
Acetyl chloride, 202
Acetylene, 28, 201
  ethyl, 207
  methyl, 205
  propyl, 208
Acrolein, 205
Adipic acid, 210
Alkali metal compounds, solid, 133
Alkali metal fluorides, 38
Alkali metal halides, 92, 174–5
  condensation energy in, 134
Alkali metal oxides, 133
Alkali metal sulfides, 133
Alkane bond energies, 146
Alkane isomers, 146
Allyl amine, 151, 206
Aluminum, bond energy data for, 47
Aluminum borohydride, 114, 183
Aluminum chlorides, 38, 94
Aluminum fluorides, 37, 94, 183
Aluminum hydride, 110, 183
Aluminum monoxide, 106, 172, 183
Aluminum tribromide, 184
Aluminum trichloride, 38, 139, 184

Aluminum trifluoride, 21, 183
Aluminum triiodide, 184
Aluminum trimethyl, dimethyl sulfide, 119
Ammonia, 21, 109, 115, 190
  atomization energy of, 21
  protonic bridging in, 114
  synthesis of, 160
Aniline, 209
Anions, nature of, 125
Anthracene, 29
Antimony, bond energy data for, 53
Antimony halides, 173, 194
Antimony hydride, 109, 194
Aromatic bond factors, 28
Aromatic bonds, 28
Arsenic, bond energy data for, 51
Arsenic fluoride, 37
Arsenic halides, 173, 194
Arsenic oxide, 35, 84, 194
Arsine, 109, 194
Atom, fundamental properties of, 5
Atomization energy, 3, 4
  of elements (table), 57

B

Barium, bond energy data for, 54
Barium fluorides, 37, 94, 177
Barium halides, 136
Barium oxide, 106, 136
Barium sulfide, 136
Benzaldehyde, 151, 210
Benzene, 29, 151, 209
Beryllium, bond energy data for, 42

215

Physical Chemistry

A Series of Monographs

Ernest M. Loebl, Editor

Department of Chemistry, Polytechnic Institute of

Brooklyn, Brooklyn, New York

Physical Chemistry

A Series of Monographs